977.8 | W9-AQM-691
c.1
Hamilton, David, 1939
June 9-
Deep river : a memoir of
a Missouri farm /
c2001.   10/02

AR   CASS COUNTY PUBLIC LIBRARY
     400 E. MECHANIC
     HARRISONVILLE, MO 64701

Deep River

Deep River

A Memoir of a Missouri Farm

David Hamilton

Around every bend the promise of the earth.
—Czeslaw Milosz

University of Missouri Press
Columbia and London

0 0022 0178707 0

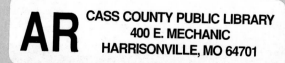

AR CASS COUNTY PUBLIC LIBRARY
400 E. MECHANIC
HARRISONVILLE, MO 64701

Copyright © 2001 by
The Curators of the University of Missouri
University of Missouri Press, Columbia, Missouri 65201
Printed and bound in the United States of America
All rights reserved
5 4 3 2 1   05 04 03 02 01

Library of Congress Cataloging-in-Publication Data

Hamilton, David, 1939 June 9–
    Deep river : a memoir of a Missouri farm / David Hamilton.
      p.   cm.
    ISBN 0-8262-1354-5 (alk. paper)
    1. Hamilton, David, 1939 June 9– .   2. Farmers—Missouri—
Biography.   3. Farm life—Missouri.   4. Missouri—Biography.
I. Title.
CT275.H2892 A3 2001
977.8'043'092—dc21
[B]   2001037679

∞™ This paper meets the requirements of the
American National Standard for Permanence of Paper
for Printed Library Materials, Z39.48, 1984.

Designer: Elizabeth K. Young
Typesetter: BOOKCOMP, Inc.
Printer and binder: Thomson-Shore, Inc.
Typefaces: Palatino, Balzano

For acknowledgments, see page 163.

For Jenny and Colin

Contents

Deep River

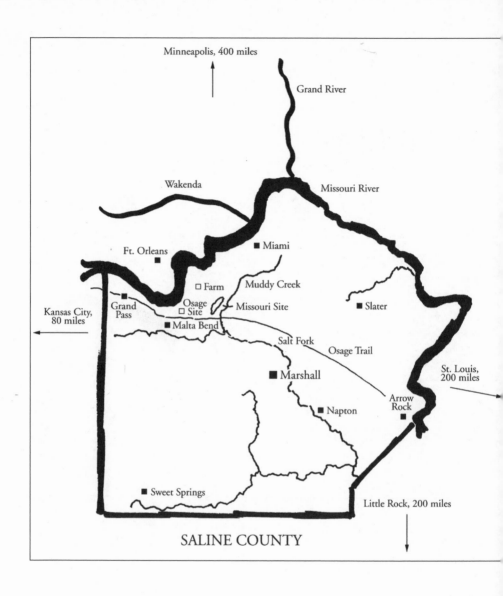

Minneapolis, 400 miles

Grand River

Wakenda

Missouri River

Ft. Orleans

Miami

Farm

Muddy Creek

Missouri Site

Slater

Osage Site

Kansas City, 80 miles

Grand Pass

Malta Bend

Salt Fork

Osage Trail

St. Louis, 200 miles

Marshall

Napton

Arrow Rock

Sweet Springs

Little Rock, 200 miles

SALINE COUNTY

Laces, an Introduction

Rivers are forever. At least we feel they should be, and I grew up beside a long one, the Missouri River, close to the stepping-off point for the Santa Fe Trail. My father and uncle cleared Missouri River bottomland, made a farm, and were working farmers. So in spite of being a white male coming through his middle years, I am, from another point of view, one of a peculiar national minority. Farmers make up less than 2 percent of our population, and those who come from farm families are not much more numerous.

I haven't lived in Missouri since I went off to college. I've lived instead in a string of university towns with interludes in Colombia and Spain. Often I've been mistaken for a city person, and my brother, who stayed to farm, knows full well that I am not of the country. Nevertheless, the country has marked me in ways that may seem slight but still catch the eyes of university friends, not a single one of whom has been from a farm.

Once in Ann Arbor, a friend and I took on some work more concrete than academic. A few peers were returning to the land, to communes and arrangements of that sort. Others were taking up crafts like carpentry. Today the escape might be to constructing websites, but then it was carpentry, and we were preparing to build shelves for a kitchen supply store. Steve, who was from Brooklyn, had talked us into the job. I could wield a hammer without choking up, rarely hit my thumb, and cut a straight line with a handsaw. Shoes were a problem, however. We would hardly appear authoritative in penny loafers.

So off to the store we went, Steve to buy work boots, I wearing an old pair but tempted to buy new ones. My boots went back to Missouri. During graduate school, I had worked summers building fences in rural Virginia. They were yellow-orange, ankle-high lace-ups, with thick cushion soles and leather strings. They needed a conditioner badly. But the treads were not entirely gone, I saw, once I took a closer look and ran my hand over them. In fact, the more I sat watching Steve try on new pairs, the better they felt. In the end, Steve bought new boots and I, feeling I owed the store a little something, contented myself with a new pair of laces. We built the shelves, and for years Steve enjoyed reminding me of my unexpected economy.

Now I don't know whether to make much or little of so small a thing. Had I become so inured to years of teasing, dismissal, or just plain lack of comprehension of the country that I shrugged off his amusement? It was good-natured, after all. Had I decided by then that the joke was on him? He's the one who spent twenty times my small investment. How much does background matter to who you are right now, and how much had it determined my choice anyway? In the world I have come to know, an indoor world of classrooms, libraries, computers, and books, a world decorated by yards and gardens but hardly determined by them, a working-class background has a certain cachet—if kept at arm's length. It denotes the liberal struggle of labor in our past. It denotes a degree of solidarity with those who fought for a share of our well-being.

But farming, in spite of the Jeffersonian visions that structured much of our nineteenth century, became a mystery by the end of the twentieth. Were farmers who also owned their land of the working or ruling class? Did it depend on whether they owned a lot or a little, and who drew that line? Farmers worked; their hands and clothes got dirty, but not all of them. What is an acre anyway? I can picture 20 or 100 or 430. But ask me to estimate my yard in a fraction of an acre and I'm baffled too.

I know I felt some pride in handling a hammer more efficiently than Steve. I'm enough of a boy forever to be glad of doing things well with my hands. Does typing count? Where then does a pianist fit in? I went off to college in that naive time, a time that must now seem absurd, when small town and rural white boys from the Midwest were seen as disadvantaged, persons to be brought in to diversify the mix. We few added flavor to the college. I worked two meals a day in the dining hall, too, and not because I had any choice. I don't for a minute equate my early discomfort with that of African, Asian, Hispanic, and Native Americans entering our mostly white colleges and universities now. I don't equate; usually their distance is far greater than any I knew. But I can empathize with them.

My friends used to tease me about the sophisticated decadence I was beginning to discover. They enjoyed finding me an innocent among them. "There goes Dave Hamilton," one declaimed over and over, "he used to be a nice guy." There was welcome in it. His tone too was good-natured. I never felt the slightest hostility and am not about to suggest that I did. But I did feel difference, a difference I've never outgrown. Nor does it run one way only. "You were the one working. Let him get out of his damn car and walk over to talk to you." That was my father's advice after he watched from a neighboring field while I stopped my tractor, climbed down, and responded to the gesture of a man in a "city car" parked on our farm road. There was a gleam in Father's eye that suggested just how much he enjoyed the thought of that man, in his loafers, walking into our field. Country people anticipate their own hostility to city airs and attitudes and are rarely at a loss for evidence.

For many years, and wearing a variety of shoes, I've kept going back to the farm. In the early nineties, we lost most of it, or to put it more accurately, my brother agreed to a forced sale. That period of conflict coincided with a time when I began to pay more attention to where I had come from and to its history. We were losing something; that was sure. High-tech farming flooded into our neighborhood as if it were a vacuum and has begun ebbing from it already—and a good thing, too!—all within a century. But farming always plows a tillage of loss.

Earlier there had been farming of another order, much of that slave-dependent. River traffic was an industry, then railroads. The Civil War grew upon border wars and those upon the skirmishes of early settlement by the English, French, and Spanish. The Missouri Indians, now entirely lost, were once the dominant tribe of the region with a city of their own on the bluffs overlooking land that became our farm. The Osage, Sauk and Meskwakie, Ioways, and Otoes each had their moments. Layers of more ancient history, sketchy but suggestive, add to a stream of incident and story, for it is not only a long but a very deep river.

And so this memoir, which is, I think, a memoir with a difference. The farm is my family's story more than my own; I cannot hold myself at its center. And that story emerges from a regional story in which loss is the common denominator. How does my own life tie into so much, and should I find my roots in all that? What kind of metaphor is "roots" anyway when movement and transplanting define so many stories? These are questions that I thought to raise, though probably not answer, about one life that could be described as laced to a farm by a river—or to a river with a farm beside it, a favored place to which I keep returning.

I
In the Bottoms

George

Long views have long moved me. Their pull goes back to very early memories, such as one of a new red tractor, in a dark, close basement, and my spending hours at its wheel, steering it into the blank of a wall that stood like all opportunity dead ahead. The Second World War had ended, and the Korean War was about to begin. Intimations of conflict were everywhere, as were memories of loss and deprivation. At eight or nine, I was too young and distant from crises to have specific knowledge of them, much less to gauge their meaning, though I heard my grandmother, with whom we lived, caution me again and again to be thankful for the food I had. And in our basement, all that winter, I harvested for others from my station on our new tractor.

Equipment was hard to come by at the end of one war and shortly before the next. My father and uncle had been fortunate to have purchased a repossessed Massey-Harris. Bright red, it had been brought into our basement until spring to protect it from the weather and to lessen the envy of neighbors. Old whitewash flaked from those walls. Hand tools and odd pieces of lumber leaned against them. One bare forty-watt bulb dangled from an overhead fixture and threw dull illumination on the wall in front of me.

It was my small theater of possibility as one winter afternoon after another, after walking home from school, I went downstairs, climbed into the tractor's seat, and steered toward that imaginary horizon. The only window was high, small, and behind me to the wintry north. A gray wash of shadow surrounded my play. But for me, that gloomy wall opened to a sunswept expanse of grain. I worked long rolling fields, under sky blue to the horizon, with wheat spreading on all sides like my expectation of summer. I thought it heroic to bring all that grain to harvest. On that tractor, I farmed more acreage than my family would ever know, and night after night, my mother had to call and call again to tell me to turn out the light and come upstairs for supper.

We were a brand-new farm family. We had moved to Missouri just the year before so my father could farm with his brother. The two men had grown up on a Missouri farm, and now in ripe middle age, they were returning to work they had known. All my knowledge of farming then came from a children's book in which the words were few and the pictures vivid as Farmer Brown honored his seasons, plowing, disking, planting, cultivating, harvesting, and plowing again. I learned to read that book before I could read, but I knew the words and so sat and turned the pages and "read" it to my younger brother, George.

As years passed, that relationship more or less reversed itself, at least with respect to farming, which is one reason why I was with George in his pickup one summer afternoon not long ago as we drove up on and turned down a levee where the ruts thinned out like time. It was a section of levee within the Grand Pass Wildlife Area that we were not supposed to be on, except perhaps on foot. Nevertheless George drove south about "half a quarter," as our uncle would have said, and stopped. No continuous forward progress was possible. Soon the tracks would end and we'd have to back out. One of nine ponds dredged in the bottoms to attract migrating waterfowl, and their hunters, lay before us; another lay off in the middle distance on my right. There was not much action in the glare of a sunny afternoon though several red-winged blackbirds held forth among the cattails. Then George glanced in his rearview mirror and said, "Here comes trouble."

Another pickup had pulled up on the levee behind us, that half a quarter back, and stopped perpendicular to the levee, blocking our retreat. One young man sat staring from the passenger's side while another sitting on a tool box in the bed leaned back against the cab. The driver got out and walked toward us.

Angular and slender, he wore boots, jeans, a short-sleeved shirt, shades, and a wide-brimmed hat. A holster and gun would have completed the outfit of a western lawman, and he was law of a sort. He worked for the Missouri Department of Conservation, which managed the wildlife area. He probably had a gun in his truck, as George has had from time to time in his glove compartment. We watched him saunter our way; not yet thirty, he could have been the son of either one of us.

"Afternoon, George. What ya-all doing down here?" The man leaned on our truck by George's window.

"Oh, just looking around. How're you, Raymond?" George looked as if winter were closing in and he knew he'd have to endure it.

"You know you're not supposed to be down here, don't you?"

"Yeah, we know. You just pull aside and we'll back on out."

There was some silence after that. Raymond's options were more numerous than the one George had suggested. George kept his seat, his hefty, work-strengthened arms resting on the wheel, his weathered face and obsidian eyes turned fully on Raymond. Perhaps his manner alone would provide the needed instruction. "All right," Raymond finally said, "but ya-all keep on out of here; we don't want any trouble." George's smile softened as he agreed, "No, we don't," and stared the agent back toward his truck. He admitted that he'd thought of several excuses, half-reasonable ones too, inspecting the levee for the levee district for example, but had thought, why fake it? Meanwhile, Raymond pulled aside, and we backed out of and off the levee with the MDC workers following behind our plume of dust to see that we got in no more trouble.

"Levee," by the way, is a curious word. From the French *lever*, it has to do with "raising" something. We levy a tax. We construct a levee, an embankment to contain a flooding river, from a tax so raised. A French monarch would hold a *levé*, or reception, on rising from his bed. All these meanings negotiate between elements not on a level, maintaining one of them at an advantage with respect to the other. At the same time the word both looks and sounds like "level," as if all were on a level, and the levee itself looks as level as a train tracing the course of a river.

A levee, then, marks a line of contact between forces we'd like to keep at ease with each other but cannot always. It can denote democracy, even egalitarian ideals, snaking through the bottoms as farmer joins farmer in their mutual struggle with the river; or it can suggest something rather more autocratic as was the case that afternoon when our neighborhood's century-long attempt to keep farmland just above river level was giving way to measures to leverage us out of the district.

For the Grand Pass Wildlife Area has taken over several former farms, all of which had been wetlands before becoming farms in this century. Of the ten thousand acres once under cultivation in the Van Meter and Tetesaw Levee Districts, almost half now belong to the wildlife area as farmland returns to managed land for the sake of wild things (and their hunters). All but a scrap of our family farm was the MDC's most recent acquisition, our land a peninsula jutting into acreage brought earlier under their control.

Our father and uncle had bought the land when George and I were boys. They bought it cheap because, if it were to be farmed, most of it had to be cleared. Finding changes advisable in their lives, they made a farm that soon supported their two families, then George's family when he joined them in the late sixties. A decade later, George had taken it all over, and it had fallen to him to oversee support of our aging parents and aunt and uncle and, later, to negotiate the sale.

From most points of view, returning bottomland to a near-wild state would seem a fine thing. George too had started as an admirer of the project as it grew up around us. But as the MDC dredged channels and ponds into the land they held and pumped river water in as refuge for migrating coots, teals, geese, and ducks, they raised the water table in our fields too, which made working them difficult. George weathered five years of faltering negotiations with persons who portrayed themselves expertly as deaf to his concerns.

When the MDC first settled in as neighbors, workers at the reserve had given George run of the place and even keys to barriers chained across its roadways. They admired his knowledge of waterfowl and his interest in all things out-of-doors. George has a degree in botany, likes to hunt ducks himself, can recognize nearly all their varieties on the wing and a whole lot of trees, weeds, shrubs, and wildflowers without having to walk up close. But as it got harder for George to farm, the MDC resisted any suggestion that their activity was affecting us. When George became convinced that the state meant to coax us beyond a statute of limitations and then have us at their mercy, he sued for their "taking" of our land without condemning it. "Inverse condemnation" is the term. A year later we settled. Now he was unwelcome on land that he had prowled for over forty years and in a position to be run off it by hirelings of a bureaucracy with which he had done battle.

These are not tragic events. In the long view, the reclamation of wetlands by the state can hardly be a bad thing no matter how clumsy or aggressive their taking. Moreover, we sold the year after the "500-year" flood of 1993. Our farm had been completely flooded and all our crops lost. Prices in the bottoms had faltered. But since the MDC wanted our peninsula badly and George had dragged them toward court, they found it harder to wait us out and so bought at a price that most would say was to our advantage.

Unhappy memories remained, though, so while I joked with George that he should set up a honky-tonk on the lousy thirty acres the MDC had not wanted and blast its sloughs with rock music on those cold autumn mornings when hunters, having just drawn assignments to ponds one through nine, snuck out before dawn to their camouflaged blinds, he daydreamed of buying more bottomland downriver and confronting the state again as a battle-hardened neighbor.

Grand Pass

Named for a narrow pass between Salt Fork and lakes in the bottoms. Old Osage Trail followed this divide, or pass.

—Place Name Study by Miss Nadine Pace

Before that afternoon on the levee, and knowing that our time was running out, I had begun returning to the farm more often. Once, driving in midwinter with the snow-covered fields shining like heavy cream in the late afternoon, and racing sundown, I swung west through broad bottoms that brought me to the river and the bluff at Waverly then east again across what was once prairie. The hazed-over sun glittered in my rearview mirror. The sky was gray white, and against the clouds I saw wisps of scriggling lines, what seemed like smoke threads blowing out and sagging back against each other, more and more of them appearing on the horizon as long strings of geese came over the rises one after another. The slower beat of their wings gave them away as geese, not ducks. They landed and wheeled up again, like a great waterspout or a slow cyclone. Thousands of snow geese, blue, white, and a milky gray. Those geese didn't insist on landing in a conservation area. They'd found fallen corn in a farmer's field and were taking advantage of it. There would be little volunteer corn in that field the following season.

My turnoff followed soon. I turned north, along a town street. "Village" is not a colloquial word here; we have small and smaller towns. "Village" has been reserved for Indian settlements even though they were often larger and inhabited much longer than any of our small towns. And of that we may be suspicious. "Village" was colloquial, at least in print, a century ago. The small newspapers of the county would mention "the Village of Grand Pass" or Miami. That no longer happens, however, and whether it ever was on the tongue I cannot say; I have never heard it.

The street ran downhill, back into the bottoms, but I went no farther than their edge then wheeled around in a U-turn and pulled up to a green sign, "Grand Pass, pop. 53." A red pickup had pulled out from a lane above and started to follow me into the bottoms. As I stopped, it paused, pulled over to the side of the road halfway down the bluff, and its lights went off. Its driver remained inside. A stranger, in an out-of-state car, I stood half a quarter away, not surprised. Nor was I alarmed, though I took for granted a gun rack and rifle. I'd given him small cause and trusted I would add nothing to that. We were long past the era of bushwhackers. I'd just come to check on something and might even touch his civic pride. Besides, as he would notice, I'm white, which is a detail I'm apt to forget since it weighs on me little. But I was multiplying my assumptions. Had I been black, he'd probably still have sat, waited, and watched me out. But probably too he'd have been a bit more on edge, as he was literally now, parked on the side of the steep bluff, the nose of his truck tilted sharply downhill, the shadows welling up from beneath it.

But it was late, and I could hardly make out the sign hanging from a small, square shelter house that roofed a pump. It was an old farmhouse pump with an iron handle you lifted and pushed down. The citizens of Grand Pass had improved their spring. A tin cup hung from the pump handle. A wooden sign, with well-crafted letters, etched but fading, read "Everlasting Spring"—

Ye Who Drink
From The Spring
Shall Return
To Drink Again.

A great many of those who did, did not. Grand Pass, named both for an upland ridge that allowed passage between Salt Fork Creek and the river bottom and for the grand passes of waterfowl that twice annually visited the lakes and marshes in the bottoms and so inspired the development of the wildlife area, was an early stop on the Santa Fe Trail. Here it angled down off the upland prairie, the "prayer" as old-timers called it. "Come in off the prayer," they said. Travelers stopped to rest their horses, their oxen, and themselves and to drink at the spring before moving back out onto the "prayer."

Crossing to the south side of the river at a ford near Arrow Rock, the Santa Fe Trail followed the old Osage Trail to Grand Pass, making a short cut of about fifty miles under the Big Bend of the Missouri on a southeast-northwest pass across Saline County, which itself is about the size of Rhode Island. After Marshall claimed the county seat in 1839, the trail angled farther south to take it in too as Marshall became the radial center and newer roads overran the old trails. But for many years the Santa Fe Trail served as a cross-county highway; farmers worked their ox-drawn wagons along it, carrying grain to a mill on Salt Fork. In a couple of days they could make the twenty or so miles from farms on the north edge of the county to the mill near Napton. There they might be stopped for a day or two waiting their turn. It was a time to catch up on what was happening in Arrow Rock, downriver near the southeast corner of the county, or Sweet Springs around to the southwest, or in Miami, Malta Bend, Laynesville, or Grand Pass, small towns strung along the river as it skirted our district.

From Grand Pass, as I walk downstream along the levee, a broad alluvial floodplain swells on my right. A corresponding bottom spreads from the left bank, though I cannot see much of it because of the timber, mostly cottonwood and willow, that lines both sides. Once glacial melt lapped up on the loess bluffs that bordered the

bottoms at widths of ten miles or more. The bluffs and the plains beyond them arose as wind lifted dust off old riverbeds and deposited
it farther away. Over eons the silt the water carried, rockless and
fertile, settled to the bottom of the river and when the river receded
became bottomland. Then during the dry seasons, some of that dirt
blew up on and added to the adjacent hills and plains.

The first white settlers carved out patches in the timber that had
once covered the bottoms. They had found their way by following
the river upstream, and they assumed, rightly, that the abundant forest proved the land fertile. They assumed wrongly that the upland
plains, with few and scattered trees, were infertile in spite of the
long prairie grass and the wildflowers. It would take a generation to
correct that error. Then farms spread all over the plains; and the bottoms, with their timber, their marshes and swags, their mosquitoes
and malaria—"autumnal fever" in early accounts—were mostly
abandoned. Only after farmers had claimed all the upland were the
bottoms reentered and cleared.

So my father and uncle participated in developing a late and by-
passed American frontier. Midcentury advances in agriculture, Cold
War competition with Soviet five-year plans, pressures of their own
midlives, and a war-bred policy of hoarding grain all combined to
spur them on. Today with more conserving attitudes in the ascen-
dant, especially toward wetlands, rivers, and streams, that bottomland is being withdrawn from agriculture about as rapidly as farmers once cleared it. It was a good farm though, fortunate land to farm,
about as good as the earth provides.

At Grand Pass, you can find a remnant of the Santa Fe Trail by
which settlers moved west seeking their advantage. A broad groove
runs through a cemetery between several old headstones and a pair
of cedars as it angles downhill toward the spring. The highway on
the south and a block of town street on the north cut off this trace of
the trail, which suggests a single giant wheel grinding and grinding
a long notch across the earth. At first one is likely to seek its mate
and wonder how wide the axle was that spanned such a giant path.
But there is no second track. Two smaller wheels on a narrow axle
cut paths side by side, and two more and two more, and over time,
wind obscured the twin depressions by blowing the dust from both
worn tracks against their eastern-most ridge. Gradually, the dust

filled back across the trail. Now it's a scrap of earth sculpture, a grass-covered slope, a long low sail bellying out from the ancient prevailing westerly winds.

The Missouri, the Osage, and many of their shadowy forebears knew of this spring and of another a few miles east where the Little Osage settled through most of the eighteenth century. Other springs pock the bluffs all around the river's arc. Water oozes from this one continuously and makes rivulets in the snow. I got out and stood up straight to acknowledge the pickup and its watchful driver. Then I stepped across a rill to look more closely at the sign. I needed to relieve myself and found a shed that I could step behind. The man in the pickup likely guessed my need as I moved out of his sight. Soon I reappeared, got out my notebook, copied down the verse, and returned to my car. As I drove uphill past him, I offered a country wave then turned down the side street to take another look at Grand Pass's single block of the Santa Fe Trail. The man in the red pickup was not so obvious as to follow me again, but he had staked out a spot by the highway, talking with a neighbor, as I drove out of town.

As I left, I remembered that I had forgotten to drink at the spring. I had though as a boy and remembered the water's mineral taste. It would seem a silly bravado to drag my escort back down into the bottoms so soon, even though an umbrella rather than a rifle rested in the gun rack in his rear window.

Sale

George framed the sale flyer and hung it in his kitchen. Auxvasse is another small country town, ninety miles east of Saline, our county. In April 1935, at the start of a new farming season, my grandparents were preparing to sell out and "walk away" from their farm. The piano must have hurt, but so too, much else. Frank, for example. Frank had been my father's mule. Frank had come as a mean mule who wouldn't allow himself to be touched, especially around the ears. By the time of the sale, my father, the younger brother, had graduated from college and was working in Chicago. But years before he had gone to work on that mule. He had framed out a pen in which Frank could not move. He spent hours stroking and whispering to Frank before he could scratch Frank's nose. It took longer to touch

PUBLIC SALE

We will sell to the highest bidder at our farm, one mile north and five miles east of Auxvasse, beginning at 10:00 a. m., on

WEDNESDAY, APRIL 3, 1935

The following property:

5   HEAD OF HORSES   5

One large bay draft mare; one yearling draft colt; one bay mare, 7 years old, in foal to draft horse; one 2-year-old saddle colt; one aged mule.

40   PURE BRED HEREFORD CATTLE   40

One Hereford bull, bred by M. B. Murry, of Hatton, Mo., registration papers will be furnished day of sale; 30 head of cows, nine with calves by sides.   All cows were eligible to registry, some are registered.

4   HEAD OF JERSEY CATTLE   4

Three head of good Jersey cows; one veal calf.

16   HEAD OF HOGS   16

One Duroc boar, 2 years old, a good one; 3 head of Duroc sows, to farrow soon; 12 shoats, weigh about 100 pounds each.

FARM IMPLEMENTS, ETC.

Eight-foot Deering binder; McCormick corn binder; manure spreader; Hoosier drill; McCormick mower; sulky rake; endgate seeder; 2-bottom 14-inch Oliver tractor plow; Janesville 12-inch gang plow; John Deere 14-inch gang plow; Janesville 16-inch sulky plow; 14-inch walking plow; 4-section smoothing harrow; harrow cart; disc harrow; lime spreader; ensilage cutter and pipe; Koger soybean huller; Case corn planter; John Deere disc cultivator; two Janesville disc cultivators; two Rankin 2-row cultivators; Janesville 6-shovel cultivator; two high wheel wagons; new grain-tight bed; low wheel wagon and frame; adjustable garden cultivator; concrete mixer; hand corn sheller; wood saw; 2-wheel power takeoff for car; Meyer deep well pump, practically new; tank heater; two double hog houses; three cattle troughs; three 50-foot rolls of picket fencing; 6-tine grapple hay fork; chain harness; leather harness; stock saddle; two brand new leather collars; International cream separator; lard press; sausage mill; small platform scales; iron wheelbarrow; 250 hedge posts; several hundred feet of narrow yellow pine flooring; 8 gallons white and light gray paint, never opened; and other articles too numerous to mention.

ABOUT 30 HENS

HOUSEHOLD GOODS

Including piano, range, rugs, sewing machine, 4-burner oil stove, large circulating heater, and numerous other articles.

TERMS:—CASH.

Ladies of the Auxvasse Christian Church will serve dinner.

G. Wilson and Henry W. Hamilton

BERNARD HARRISON and FRED RENTSCHLER, Auctioneers.   MILLER GALWITH, Clerk.

AUXVASSE REVIEW PRINT

his ears. Father assumes the former owner had tried to control Frank by putting a twitch on his ears. That would be a band around them with a stick to twist it. Eventually Father got to Frank's ears and then got a bit in his mouth and Frank became "a good mule" though "aged" by the time of the sale. Among other things, my father had proved to be a "mule whisperer."

The disc cultivators had two settings. First the two discs would be placed apart, at almost the width of two rows, so they could throw dirt from beside the young corn to the center of the row, retarding more than killing weeds growing there. Then, when the corn had got its start, they would be reset close together to chew up that center and toss loosened dirt back toward the base of the corn. The cultivator could only work a single space between two rows. Cultivator shoes, a later development, were staggered to scratch up that same width, both close to the base of plants and in the middle between the rows. That meant one less operation before "laying by" the corn, which meant getting the corn well enough started and ahead of most weeds to grow on its own toward harvest and to trust in the weather for the rest. Frank or one of the draft mares would draw these cultivators, as they drew most of the equipment.

My grandfather and grandmother were selling out and moving near my uncle, who was assuming the role of patriarch. Grandfather had sold a hundred head of cattle and had taken the money into the bank late on a Friday afternoon. His loan was due. The banker accepted his money for deposit but said it was too late in the day to process paying off the loan and to come back early Monday morning. Over the weekend the bank closed. My grandparents' money went toward paying off the bank's creditors, and they eventually got back ten cents on the dollar. Their loan, however, remained due, which makes this an emblematic story of the Great Depression. Eventually, the banker went to jail, but that was the end of Grandfather's effort to farm. He would turn sixty-six that fall, though no one said in those days that at sixty-five one could retire, and he lived another ten years, always a bit ashamed of being a burden on his sons.

Almost a decade before the farm sale, my Uncle Henry had become the agricultural extension agent of Saline County and had made his home there. Another decade would pass before a 430–acre farm in the west bottoms became available, fewer than 100 acres of which had been cleared. Our two families bought it, and we moved to Marshall, Missouri. My mother was from Oak Park, and all through the thirties and most of the forties my father had worked in Chicago. My grandmother had also been from Chicago. For two successive generations, in my family, Chicago women found their way to Missouri farm towns, which is a subject to which I must return.

Buying and clearing bottomland meant an invigorating midlife change for my father and uncle, who found a way to "go back to their roots," as the next generation would say, or at least back toward them with the healthy sweat of hands-on labor. More than once, my father has said that *Walden* was the book that most influenced him. But farming was also a practical choice. With a farm, they just might pull the older generation through and give the younger one a start.

Underlined Passages in My Father's Walden

> The mass of men lead lives of quiet desperation. What is called resignation is confirmed desperation.

> To be a philosopher . . . is to solve some of the problems of life, not only theoretically, but practically.

> Shall we always study to obtain more of these things, and not sometimes to be content with less?

> A man is rich in proportion to the number of things which he can afford to let alone.

So far, Father's selections are about what one would expect, and I've long assumed that the second of them was the most influential. But then comes the less expected poetry of a fifth passage:

> It is well to have some water in your neighbourhood, to give buoyancy to and float the earth.

When, in the mid-thirties, while working in Chicago, my father underlined that last passage, he could not have foreseen that he would become a river bottom farmer and learn a great deal about a river's way of giving buoyancy to and floating the earth, not to mention floating away many who had once lived alongside it.

The Bottoms

To hit bottom. The bottom rung, rock bottom. To scrape the bottom of the barrel. If there's a top of the morning, there must be a bottom. Lake bed but river bottom, though there's a riverbed too, at the bottom of the river. "Such a dream I had," says Bottom the Weaver. "It shall be called 'Bottom's Dream,' because it hath no bottom."

The river flows above its bottom, which can run shallow or deep, just as the bottom of a ship, the portion below the waterline, can glide just beneath the surface or cut down deeper. Ships too are bottoms, as in "a fleet of forty bottoms." To say "that horse has a good bottom" means that it has a strong formation and does not tire easily.

By a geological clock, the land is flowing. Sixty-five feet of it cover a limestone bed, and we call the top of that the bottoms—all the river's adjacent floodplain subdivided to plural form by the divisions of communities and circumstance of the people who live along the river. As early as 1819, an expeditionary report by a soldier under the command of Maj. S. H. Long remarked on the "distinctive names" already attached to settlements of the area. "We pass on the south the Chenai au Barre, Tabeau, Titesaw, and Miami bottoms." The Tetesaw Bottom, variously spelled Petitesas and Teteseau and pronounced "teetsaw," became the location of our farm.

The name is also given to the adjoining plain. One explanation finds it sloughed off from "Little Osage," from "Wazhazhe" or "Wacace," two ways of transcribing the name by which those people appeared to call themselves. One can imagine "Petite Wazhazhe," "Wasazs," "Sas," and "Saw" on the tongues of French voyagers and trappers. Another is "Plain des Petites Sauts," which would have derived from the numerous falls running off the bluffs that divide plain from bottom and that feed the springs. Yet another hinted at in "Teteseau" is little "kettle" or "bucket" for the ancient lake bed higher than the river bottom that became, in time, the prairie. So some have thought the French meant to say, "little cradle of the prairie," without trying to explain their translation, but "Little cradle [or kettle or bucket] of the Little Osage Prairie from which small falls leap down" ought to encompass the idea.

Well below that cradle, the bottoms make up a floodplain roughly level with the riverbank. It is an alluvial plain formed from sediment.

At times the Missouri filled the whole of the plain as glaciers melted and sent rushing back to the sea the considerable fraction of the earth's water that had been locked up in ice. Sediment in the water sank slowly to the bottom. Then as the Missouri quieted gradually to the river we know, it wandered back and forth across the plain and kept on filling in those bottoms.

The river cut a channel across the bottoms. The river would knife along like a plow carving a furrow into the land. Plowshares are also "bottoms," with a farmer working a single-, two-, or perhaps a four-bottom plow. Prevailing winds would urge the river more toward one containing bluff line than another. Then the winds would shift and urge it back. Or once pushed and cut deeper, another spring melt, with new rains and resulting flood, would swirl against a bank, catch on a cottonwood rooted in it, turn, and cut a new path across a bend, reforming the channel. Or sediment building up would turn rushing water away from a bench or table it had formed. Hence the rises in the bottoms, the "eminences."

Since we are a bookish people, it didn't take long for someone to imagine the river writing and rewriting its way across the land. Mark Twain made much of learning "to read" the neighboring Mississippi. Similarly, Thoreau, early in *A Week on the Concord and Merrimack Rivers*, celebrates his farmer neighbors as writers.

> You shall see men you never heard of before. . . . Look at their fields, and imagine what they might write, if ever they should put pen to paper. Or what they have not written on the face of the earth already, clearing, and burning, and scratching, and harrowing, and plough-ing, and subsoiling, in and in, and out and out, and over and over, again and again, erasing what they had already written for want of parchment.

The figure has a long history. A chorus in *Antigone* chants its equal.

Early writing was lapidary, engraved in stone. The river, too, first carved its channel through limestone bedrock then carried in its fill of suspended, gradually settling earth. Over time, soil the river washed along settled somewhere, then often somewhere else. Soil and sometimes sand. As men and women began to map the region, the river did not always mind their maps. Parts of Carroll County to the north of the river became Saline on the south, and parts of Saline County became Carroll. A channel would shift from the right side of an island to the left, with the right side filling in until it became continuous with the bank. So we came to farm accreted land.

Much earlier, the Missouri Indians had become familiar with and filled in the land, a small piece of which became our farm. Their nearby townsite covered the upland ridges, and they gardened and hunted bottomland on both sides of the river. Meanwhile their cousins, the Ioways, the Otoes, and the Osage, all Siouan language speakers, filled in to the north, west, and south. The Otoes may have entered the region with the Missouris (or Missouria as was once said) and divided from them, moving west, then north until they came to "Ni-bra-th-ka, with a trill over the 'r,' meaning flat or shallow water." So they named Nebraska for the Platte, "a mile wide and foot deep." When the Sauk and Meskwakie pushed and were pushed into the region, they contested with all four.

These nations, you could say, taking a geological view, were more sediment that built up, settled, shifted, and defended their sites of tentative advantage—as white settlers would become later. The primary sediment though, in granular form, was sand or silt or clay, with sand the most coarse and clay the finest. Sand alone isn't much to farm. Clay, though very fine in its granular form, bricks up hard as it dries and is not easy ground to work. Silty loam has more organic matter, more humus, and more sand. Our Front Field, the one bordering a county road, was mostly sandy loam. It worked easily and was less lumpy beneath your feet. If you knelt and took a clod up in your hand, you could work it apart in a moment into potting soil.

At the other end of the farm, our North Field was gumbo, a dark clay soil that is unusually sticky when wet. Sink in to your knee and it will suck your boot right off. Common through the lower Missouri and Mississippi valleys, it gave its name to the soup, or the soup to it. The first definition in my dictionary is for the gummy pods of okra that give that soup its body, and the word is said to be of African origin, from the Bantu. My father loves okra; we had plenty in our garden. As soil, gumbo has more clay than silt and bricks up too leaving deep cracks in the dark earth as the summer lengthens. It was the hardest ground to work when wet, the least tractable. When the plow cracks gumbo, the earth falls away in chunks or rolls over in long, thick sheets like a quilt too thick to fold and put away.

Bottom farms don't require top farms for the sake of definition, and in fact some of the county's top farms are in the bottoms. But bottom farms do require, for the sake of definition, farms in the upland plains or on the hills. The upland Tetesaw Plain overlooks the Tetesaw Bottom. The plain, a combination of alluvium and loess, is virtually rockless and deep. The bluffs on both sides of the river, the plain itself, and the soft defining hills around it are all mantled in loess, which means that for ages we can count but not comprehend the air was often filled with earth in its finest form. For a couple of million years the soil kept building—some washed in by the river, some blown in on the wind. Farmers on the higher plain are thought by those in the bottoms to be looking down upon them.

In 1952, when levees had yet to be repaired after a major flood of the year before and so the bottoms flooded for the third time in our first five years of farming, my brother and I assured our parents that they would not have to maintain our fifty-cent weekly allowances. We stood on the bluff with upland neighbors and looked out over an inland sea that covered everything for miles in all directions except behind us. It would be another decade before I saw an ocean. We figured we could mow lawns and rake leaves for our pocket money.

"We were moving peacefully in clear, calm water," Marquette recorded in June 1673, as he and Joliet descended the Mississippi,

> when we heard the noise of a rapids into which we were about to fall. I have never seen anything so frightening, a confusion of entire huge trees, of branches, of floating clods of earth, rushed from the mouth of the Pekitanoui [Missouri] River with such force that we could not pass without grave danger. The agitation was such that the water was all muddy and could never be clean.

June is flood time and "a June rise" part of parlance. The explorers had only happened upon the usual. Early newspapers record flooding for almost as many years as not. In 1921, for example, "all the pecan groves in the lower bottoms were flooded and water drove the Chevalier family to seek shelter upstairs."

Later, we became the Chevaliers' immediate neighbors. Joe, the youngest son, my age, often worked a field across the road and took more chances with his tractor than I did with ours. One spring day he ventured one round too far into a swag and the tractor sank to the axle on his large back wheels. There he sat in the distance, spinning and spinning and watching his wheels as if studying how far down he could dig. "I've heard of China," he might have been thinking. We found a log chain, hitched our tractor to his, and pulled him out. Turning to farming after the eighth grade, Joe stayed in the bottoms. During the 1993 flood, he kept his outboard "loaded flat," bringing his gunnels down to water with the last possible sandbag, and putt-putted out along the levee to where they'd best be laid. He worked as if all the river were but another field to him.

In a letter to my aunt, my uncle describes the 1943 flood—only a moderate one by his records. Coming into one farmhouse by boat with neighbors, he found a cat "dry but weak" in the kitchen. Unc took him upstairs, set out bowls of canned milk and water and opened a loaf of bread. The cat could make it several days more "if necessary." In a second house, the men "chinned" the water as they moved clothing and a mattress upstairs. As they wrestled the mattress up and threw it on the floor, a black snake poked out his head. Downstairs,

> just about everything was floating but the stove. A washtub was there, dry inside, I threw the pillow in that and let it float. I took off the lamp chimney and put it on a high shelf, so that if it went much farther the chimney would not stop the upward progress of the table.

That's the uncle I remember, improvising like a sergeant major, using what was handy to keep a pillow dry and foreseeing that the lamp chimney, catching first against the ceiling, would shatter rather than hold a table down under water determined to rise. And so it went in 1947, '51, and '52 until the Corps of Engineers stabilized the river channel, raised levees, and built a series of dams upstream. Then the floods were fought and "won" for forty years.

To say this work was controversial would be an understatement. The levees controlled many a June rise but brought flooding, when it did occur, to new heights by containing and so increasing the speed and force of the flow for as long as possible. Thousands of acres on upstream tributaries were lost to lakes behind containment dams in what seemed a system of denial to some, including Native Americans on their reservations, for the sake of barge traffic and more dependable farming elsewhere. The river served as an auxiliary rail-road, and the actual railroads took quick advantage of more dependable, better controlled bottomland by laying hundreds of miles of track along a level surface. Cold War planning for better lines of internal communication was a factor. A grain surplus was an advantage. Big-time agriculture was never more positively envisioned, and big-time vacation lands began to beribbon the Dakotas.

The great flood of 1993 came on slowly. There was the usual runoff of winter melt from the northern forests and the western mountains. Then came the rains. Spring rains turned into summer rains. All June it rained. Every other day it seemed to rain, which is probably an exaggeration. No, every day it rained somewhere on the river system. The rivers swelled and kept on swelling while rain rained on itself. Fields lay patched then smothered in ground water that could find no place to soak in. Every stream, creek, and rill became a river. Creek beds that one could have stepped across filled and rose as torrents. Satellite photos centered on Iowa showed that its density of water approached that of the Great Lakes. All the rivers flooded. But rather than cresting and subsiding, they were refilled by new rains, sending crest after crest rolling down each stream. All along the rivers, levee systems held and held and then were overwhelmed. With '51 in mind and so with only slight exaggeration, the '93 flood became spoken of as a fifty-year flood. As further crests occurred and more levees broke, it became a one hundred–year flood. Before it was over, we spoke of a five hundred–year flood as if we all had personal recollections of pre-Columbian eras, and we watched water fill the landscape as must have been normal toward the end of a glacial period when the bottoms were smoothed to ten- and twelve-mile widths by long-term glacial melt. Driving through the bottoms later, George and I found watermarks on trees two feet above the top of his pickup cab.

When my father and uncle came to the bottoms, they bought the land that was available and that they could afford. If they wasted any time wishing they farmed on the Tetesaw Plain, from the edge of which we looked down on one flood after another, I failed to hear about it. Much later, with my uncle dead and my father retired, and having sold 400 acres of our 430 acre farm, we heard of an upland farm for sale and went to look it over. It was rolling, with creeks and draws working their way toward the river. It had a pond fringed with trees. I couldn't tell whether my father coveted the place, high over the floodplain, or if he just wanted to stick with farming. "We know farming," he said. His father and his grandfather had known farming. Like my brother, both had been Georges, named for those who work the earth.

At farming we were competent; that's what my uncle would have said. "Competent" was his favorite word of praise. "He's competent," he would say of another man, likely a neighbor farmer. When I finally looked it up, I found the Latin *petere*, "to compete," "to strive." A competent person strives capably against others. The root, moreover, that small part, *pet*, names feathers. Swifter than it seems, "competence" is winged and flies, or flows, along the beds of ancient rivers. So if you get right down to the bottom of it, my uncle's wry understatements could be lost on the ignorant.

Clearing

Notwithstanding the general prairie character of the county, a vast belt of timber land, of from one-half to six miles wide, fringes the Missouri river, and corresponding belts fringe all the lesser streams, of which there are many. . . . In the fall of the year the vast wooded bottoms . . . abound in pecans, hickorynuts, and "mast" of nearly every kind. Wild grapes, summer grapes and fox grapes, flourish and bear luxuriantly.

—History of Saline County, Missouri, *1881*

When my father and uncle were clearing, I tagged along whenever I could, which has much to do with my intermittent absorption in outdoor things and everything with my learning to use an ax. For in a frontier community such as I could pretend ours was, receiving an ax was a crucial step toward becoming a man. I had a camper's ax, more than twice a hatchet but not full-sized, and was learning to use it. Clearing our farm, my father and uncle used bulldozers, chain saws, dynamite, and fire, but they also used axes. Several well-worn, double-bitted axes were among our common tools along with a handful of heavy files to sharpen them.

Jess Miller, who worked for us, was a wizard with an ax, and I saw evidence of it. In a kind of John Henry contest, my father and uncle raced him one day, they with a two-man chain saw, he with an ax, to prove to Jess that the saw was more competent. They managed, though the awkward two-man saw could not do some things as deftly as an ax, and it took them most of the day to establish a lead Jess could not hope to overcome. Once or twice the chain broke and had to be repaired. They used up three such saws clearing the farm.

One of my father's claims about Jess was that he could take a tree leaning one way and drop it the other. As I remember now, the feat had something to do with a particular tree being festooned with wild grapevine and his using that tangle and the tree's own weight as leverage. In *Two Little Savages*, by Ernest Thompson Seton, the story of rural youths who play Indian and learn woodcraft through a coming-of-age summer, and much my favorite book at that time, one protagonist, named Sam, takes up a bet that he cannot fell a tree against its lean. A two-minute time limit figures in the story. But Sam does, and Seton describes in detail his first cut and his second, then how he gets the tree to break backward, you might say, and waits tantalizingly for just the right wind to give it an extra push so that, with his last strokes, he fells the tree, and it drives a stake into the ground exactly along the line prescribed.

My own axmanship was less competent. One of our jobs each spring was to walk through new fields and cut out willow and box elder shoots emerging from land we had cleared and plowed the year before. The old woods persisted in the young shoots. Those would be limber, sprouting up crookedly from uneven soil that had been loosened when turned over and then had settled through the winter. The shoot was light enough to cut in a stroke, but only a stroke well placed. You had to strike just beneath ground level so your ax would meet resistance. And you had to read the shoot into the soil to know the angle at which to hit. If you struck too high or missed its curve into the ground, the shoot would quake, and a small shock reverberate along the ax handle into your hand and arm, and you'd break stride and have to try again. Part of the skill entailed moving through the field steadily, sizing up where and how to hit by instinct, much as a boy running along a path through the woods shuffles his stride without conscious thought to better leap over a log. The unconscious element was critical. Having a sense of what lay *sub-tela*, beneath the weaving of the soil, was subtle.

Much the same was true of splitting logs. Far better to guide the ax deftly than to whale away with it. Crucial to success were intuiting where to put the bit against the log and then placing two strokes on the same line without thinking too much about it. It also meant keeping your ax clean and sharp and knowing how to use the file, stroking evenly toward a center point along the line of the handle. Work like that partook of romance for me, though only because I never had to do it on a daily basis. Even now I lament storm damage less than I should. It justifies getting out my adult-sized double-bitted ax that I bought for three bucks at a garage sale and restored using oil, a whetstone, and a heavy file that I bought for the purpose and that cost four times as much as the ax. The handle was split too, and I set in a new one. Now it clips off the limbs and divides logs for the fire.

Felling trees, though, was just the glamorous part. My father and uncle bought a secondhand John Deere D with cleated, iron wheels—slow, simple, and powerful for its time. They bought a Model A Ford and used that as their first farm car. Hitching the Model A to a wagon, they hauled to the farm an iron V that they had had welded together from two side frames of a car that had been junked. It lay on the ground like a twelve foot tuning fork and measured eight feet across the open end. They were tuning up their land, and they needed more than saws and axes. They needed a rake, the rest of which they made from cottonwood two by tens and four by fours sawn from oak.

They assembled it in the shade of cottonwoods on a house site abandoned before they bought their uncleared land. Setting upright a dozen four by fours, they bolted them to an axle joining the open ends of the fork. These were to be the rake's teeth, but something had to hold the teeth to the ground. So they bolted four lighter planks at 90-degree angles to the teeth of the rake. Then they hinged a long, running two by ten to the iron frame and held it back with a spring over the ends of those horizontal planks. The plank kept the teeth of the rake in place, dragging rather than turning while being pulled forward. The teeth then pushed brush into a windrow.

When enough had bunched up, the two by ten could be pulled forward against the spring and the horizontal stays released so that the teeth fell backward and the axle turned. The teeth would pivot

180 degrees until their formerly upright ends fell straight to the ground. Turning, they left the gathered brush behind. Another set of stays then reached forward on the horizontal and the two by ten sprang back over them to steady the teeth, which could then start another windrow.

Jess had done his part first, as had a bulldozer with a shaving blade. After taking a chain saw to the larger cottonwoods, they pulled out most of the stumps with a tractor and log chain. The toughest required dynamite. The rake came into play when they began to see a field. One wonders how the pioneers managed, without the oldest tractor I've ever driven, much less the bulldozer, chain saw, log chain, dynamite, and the ingenuity of a couple of vigorous and experienced men, the younger one, my father, trained as a mechanical engineer. They both remembered smaller, lighter, mule-drawn rakes that had made windrows of cut hay.

Their implement was typical of their faith in the makeshift, in nothing's being junk that could be salvaged for a purpose, in always finding a way to invest more of themselves since that's what they most had to spend. My father learned to weld; the nearest blacksmith was miles away, which meant a day lost to the smallest repair. A pile of scrapped equipment mushroomed beside the machine shed, and they always searched it first to construct a needed part. It was a pile of toys no one ever made them put away. Often it proved essential.

What fully prompted their labor, I cannot know. Years earlier, on only their second anniversary, my father and mother had agreed "to slow down and enjoy life" if Father wasn't a business success in ten years. It didn't take quite that long, for eight years later, stuck in middle management, he became restless. Uncle Henry was older and had long worked around farming though not at it. Perhaps he had assumed he would take over the "homeplace" near Auxvasse, which his parents had lost. To begin to recover that loss, he bought 80 acres of uncleared bottom land and started again. On vacations, Father came from Chicago to help. Then the 430 acres became available, and they seized the chance, investing all they had and all they could borrow.

Years before, Uncle Henry had done an unusual thing that may have reinforced his desire. Shortly after the First World War, he had worked in Europe for Quaker relief missions. Through the long winter of 1923–1924, on the heels of famine, he had been stationed on the Polish-Russian border, between Vilnius and Riga, and had directed efforts to rebuild villages and reestablish farms in territory that had been systematically depopulated with most villages burned to the ground. But the refugees were returning, living in hovels and burrows, and perched on the edge of starvation.

Uncle Henry worked with peasants and with veterans of the Polish and Russian armies. He started a school for orphaned boys and helped train them to farm. When his duty ended, he traveled to Moscow and decided against standing in line to see Lenin's body, the line being "too long." He worked his way south through Kuybyshev and around the eastern side of the Aral Sea all the way to Tashkent at the foot of the Himalayas. In one two-week period, he traveled over six thousand miles by Russian railroad, always in third class, the level just above bare boxcar, and only took his boots off twice.

The year before had seen the worst of the famine. Writing from a town on a tributary of the Volga, he told his mother that "for two winters the wagons have picked up the dead. Men have eaten dogs and dogs have eaten men." In morning sun, he saw the pock marks on walls "made by the rifle fire of passing revolutionary and counter-revolutionary armies." He photographed the burial grounds of German, Polish, White Russian, and Bolshevik armies and listened to witnesses from all sides tell their stories. He became cautiously friendly with the secret service agent assigned to follow him between Moscow and Tashkent and heard the agent express an interest in coming to America. Would there be farmland? He knew at close second hand the fragile and contingent nature of supplies and the vagaries of weather, all of which was reinforced a decade later by drought and dustbowl and the Great Depression.

He found beauty too. Writing in his later years he described "certain sounds that will live always in our memories" and ended his Russian memoir this way:

> It may be the moaning of the wind around the corners of the house on a cold snow-covered night as you pull the covers closer about you, firm in the knowledge that out in the barn the livestock in your keeping has been cared for.

Perhaps it is the swirl of water as the ship cuts through the black and silent sea and the stars shine upon you. The smell of salt water is in the air, the bell on the bridge tolls the hours of the watch as you and other members of the crew sit or lie about the deck and smoke a last pipe before going to your bunks.

But the most vivid sound of all to me is the creak of saddle leather and the steady muffled thud of horses' hooves in the snow, in a silent and very distant Russian forest.

The first memory comes from his youth in Missouri and from the farm he longed to recover. I recognize as true to him his taking responsibility for the livestock. The second recalls his working return from Athens to Alexandria to home, a rare interlude at sea for a farmer. The third fixes that formative winter when he helped a rural people loosen the grip of famine. He was twenty-five and in charge of veterans at a work station on the Russian-Polish border. Its essential sounds echoed the knowledge of his youth and were translated over the next sixty years as the sounds of farming. In his younger brother he found a willing partner.

To Clear

To clear one's mind, or heart, or throat. To be freed from burden or obligation or from legal charge. To gain or profit by a knowable amount. To settle things, to clear them up. To own something "free and clear," neither mortgaged nor encumbered.

To come into a clearing, where the sun shines. For the extent of the clearing to be without obstruction to your view. To be free from confusion and so free of doubt. For a solution to be evident to the mind and a problem, therefore, solved.

When my father and uncle cleared land for a farm, they engaged in a foundational act. Since the eastern half of the United States had been virtually all forest when Europeans arrived, almost all cities and farms unto the Mississippi began with axes ringing. "We are here," they sang, "making it clear that we claim a place for ourselves." It's hard to lament that overmuch when here we are—cities, commerce and industry, universities, and farms.

West of the Mississippi, as woodlands gave way to prairie and plain, much less clearing was required. Instead we think of our pioneers as sodbusters, turning long grass under and building sod houses. They cut what they found before them and made their first homes out of the material at hand. In Saline County, especially on the Tetesaw Plain, there were few trees, and settlers turned farms up out of the soil without clearing much forest first. But the bottoms were heavily wooded. To farm the bottoms, timber had to be cleared. Trees, shrubs, vines, and brush.

In its unflagging attempt to clear more things up, my dictionary says that the Indo-European root of "clear" leads both to "clamor" and to "claim." Clearing the land set up a roar. Axes, bulldozers, chain saws, and dynamite, what a ruckus we made. The oldest plat maps show all the county divided up and claimed by someone. But much of that was in timber, especially in the bottoms, and until an ambitious farmer came along to clear it, it may have been a deed on paper, but it was much less a deed in fact.

As my father and uncle cleared their land, they began to see a farm. They could see ahead to planting and harvest because they had gained a clearer view of what they possessed, by deed, surely enough, but also, as another idiom goes, by making the fields their own. They could look across a would-be field and begin to see its shape and condition. The geography of the farm became featured by what they saw: the Strip of Timber, the New Field, the Half Circle of Timber, the Long, Narrow Field, the Dogwood Patch. I always guessed about the Dogwood Patch. I didn't help clear it, and then it was gone; and so I have always been uncertain of its place except to know that it named the southeast corner of our one hundred–acre Middle Field. Its features lay in the memory of my father and uncle rather than in my own, and it must have been a difficult piece of ground since they referred often to breakdowns on it.

My father and uncle cleared the land and made a farm from it. Later they cleared off all the debts they had incurred in its making. Thus they cleared the way for another generation. They got a lot out of their fields. They wouldn't have responded readily to "What's your field?"—the first question at a university. They could better answer what was in one—corn this year, beans last. Perhaps their usage foreshadowed some tactful doubt about possession, even though the deed was recorded and money laid down on the marble counter in the bank.

Once a New England friend complained that midwesterners fail to hunger for "peaks of excellence." Our accomplishments tend toward the horizontal. Farmers bend to their work on both sides of the road, and river, and in the long run, health, weather, and perseverance level out much difference. I find it hard to imagine "He's an eminence in his field" being said of a farmer. On my uncle's tongue an "eminence" was a visible rise in the lay of the land, a bench or table. It might swell up waist high across a field. Your tractor would chug a little more crawling up on it. If my uncle were to use "eminence" to describe a neighbor farmer, he would have meant only that through sustained competence, over many years, a man had pulled himself to some small advantage, which is exactly what he and my father managed. Later George joined them at it.

Which makes none of us clairvoyant, another relative of "to clear," though chiaroscuro effects they knew well: the play of light and shade all over the field as cloud shadows cross it; the interplay of willows along the drainage ditches, and through the swags, with corn in the well-cleared field. Or of plan with execution, of the present with the past, of the generational and social demands made, over time, on a piece of land. Of a farm at cross-purposes with wetlands, and of the Missouri Department of Conservation with us. It is probably an error to think that we ever clear things up. At best we clear the way for a spurt or two of what looks like progress, most of which calls for correction later on.

Winter Onions

My uncle loved winter onions. He planted them in his backyard, beside our garage, and behind the first rough-milled machine shed out at the farm. Even now, along the old Van Meter Levee, abandoned half a century since, bordering that last strip of timber we never cleared between the Front and New Fields, you might be walking and come across another patch of winter onions. George found a bunch in his garden, high on the loess bluffs overlooking the Missouri. He found them setting out new onions two years after Unc had passed beyond their harvest.

George cut a bunch and took them to the cemetery. Each nubbin was drawn tight at its root knot onto the green stem. George separated them carefully, their delicate, brownish shells fraying. Rose tints peeked through the fragile skin. From the tips, tiny green tongues of new onion stems were forming. He planted them beside Unc's headstone, root knots down and spreading, from which start they "walked" all around the stone. Now a decade and a half later they still thrust long green stems out of the ground each year and set new onions. "You'll find it easy," George said, speaking of the gravesite. Onions stand out well among plastic flowers and chrysanthemums.

After finding those onions and visiting Unc's grave, I went to see Aunt Jean, who had just moved into a retirement home. She was sorting through another box of papers she had brought with her. A dozen flat copper replicas of Spiro Mound effigies glittered on the wall behind her. She and a friend had made them, indenting copper sheeting with the sharpened tine of a deer antler. For over forty years, I'd eyed them on every visit to her home. A grandfather clock surprised me with its ring right on the quarter hour.

"That's good, that's good," she said when I mentioned what George had done. Her hair was drawn up into a bun almost as tight as the tiny onions. Her glasses set off her face with their wire frames. Though she was very thin, her embrace was strong while she laughed at how George had remembered her husband.

I remembered him too, and surprisingly often, as for example in his pickup on gravel roads in the county. By the early 1970s, he'd known Saline County farmers for almost half a century and had become one. His pickup had no gun rack. I can't remember him ever with a gun. He had neither taste nor time for hunting. A log chain lay in the truck bed alongside a double-bitted ax, a tool box, a couple of rough-milled cottonwood planks splotched with oil and dust from country roads, a grease gun, and a hoe. Two pipes lay stashed in the glove compartment along with Prince Albert in his can and several cellophane-wrapped cigars. Two five-gallon army surplus gas cans filled for tractors rocked but stood upright in the back bed, a gallon glass jug beside them, wrapped in burlap to keep drinking water cool for the fields. Spot, a dog who'd been left behind by a farmhand when the water was rising, rode with us. Unc had turned back, found and picked up Spot, and had kept him.

Unc had a pipe in his mouth, an old leather tobacco pouch visible in his denim shirt pocket, and a box of kitchen matches half open on the dash. His tanned forearms rippled as he gripped the wheel, his thumbs splayed upward flat against its curve, reminding me of how he fought the roads back in 1951, our first big flood year, when the roads were more mud than gravel and his vehicle was our Model A, with lots of clearance, narrow, digging wheels, and the ability to slither through whatever roads were not yet under water. We had fishtailed in and out of ruts, and sometimes Unc had got both his elbows and his chest clamped down on the wheel to control the slides. But this was a dry summer on much better roads, and he could drive and talk.

Each turn and farmhouse prompted another story. Something Sam Irvine had said about Malta Bend on the bluff above a bend named for the *Malta*, which had sunk there in 1841. According to newspaper reports, the *Malta* had carried furs, although some said it was loaded down with "a lotta malt," which residents recovered and drank. Irvine, whose family predated us in the county by about a century, claimed it carried wool army coats intended for an outpost upriver and that local boys salvaged as many as they could and wore them around the Bend for many winters thereafter.

Or something about Laynesville, "lying yonder" under water. In the 1870s, it had been a hemp shipping center. John Layne laid it out, back when the primary ambition had been to establish a town, not just begin a farm. He had hoped to edge out Miami. The flood of 1903 washed away all that was left of it, leaving less than could be found of a burned Polish village. Long before that, though, and before the Civil War, or the War Between the States, as my uncle but not my father called it, a Benjamin Hinton had been murdered near Laynesville and a black man found with some money and lynched for it.

"See how the river swings west and north ahead of us, round what we call 'the island'? Old plat maps show that it once zig-zagged east and cut right across our farm. Accreted land. We farm accreted land. Once it was riverbed, then later on the south edge of Carroll County rather than in northwest Saline. That's Jack Little's house. We bought from six Littles, one was married to a Haynie. Only the Front and most of the Middle Field were cleared, less than a 100 acres out of

430. That was one hot, dry summer. Soon after we bought the land, Jack tossed a match into the brush, said he was doing us a favor.

"Don't suppose Gussie thought much of that." Gussie Chevalier lived just east of us with all his farming sons. "You remember how Bud tried to get in one round too many one spring and lightning got him instead? Hit his rig grounded by the rotary hoe and busted out both those big, liquid-filled back tires—broke his shoelaces too. Joe's about your age, isn't he? He quit school but he knows the river.

"We've got three more Oneota sites here west of Van Meter, either in the bottoms or on the slope down from the Tetesaw Plain. There're a bunch more in the county, at least twelve we know of. Did I ever send you a copy of my report on the tobacco pipes of the Missourah. Good, I'm glad. You always fancied that smooth red pipestone. Ever find a pipe yourself? There was a time when most of the farmers in the county were lookin' for me. They'd turn up stuff and let me know. Found a bunch of pipes. Those fellas must have had the habit.

"Over there on the Pinnacles lived ol' Jim Thorp. He carried mail out of Miamah and worked snag boats on the river. Whole trees washed out of the banks, roots an' all. Snags sunk many a steamboat. Thorp never married. Died here awhile back and kept a scrapbook on Miamah almost till he did. Eight volumes of old photographs, newspaper clippings, and his summaries. Your Aunt Jean got him to give it to the State Historical Society. She's on the Board you know. Don't think Jim ever graduated, but he cared a lot about Miamah and its schools.

"He liked to tell of the old high school burning down, how he stood and watched it in the rain that night. Not that he wept! All but part of the west wall went. That was the year after the flood washed Laynesville away. Thorp reports that the new school cost under $7,000. 'Pro Populo 1904,' it says above the door. You know what that means? They hung Shakespeare and Jefferson on smart medallion busts, looking south; don't even get to see the river, just kids comin' in the door. Now tell me, Dave, what is it that you do?"

Unc knew well enough. My Ph.D. was recent, and I taught English literature at a university. What was I teaching? Chaucer, my favorite course was Chaucer. "So what exactly do you do when you teach Chaucer?" We swung through another section-marking curve, almost a right angle on the road.

"Well, we read his tales and discuss them. I explain what needs explaining." Unc's pause gave me a little more rope, so I kept on falling. "We interpret Chaucer's poems. I help the students see what Chaucer meant."

There was another moment's silence as we passed another farm, which might have prompted another tale. Instead Unc asked, in his voice of gravel, "Can't you let Chaucer speak for himself?"

"I don't know," Aunt Jean said, bringing me back to George this time, "I don't know what we'd do without him." I don't know what I would do without him either, he who stepped in for me and shouldered farming. Our parents wrote out several other goals on that napkin during their second anniversary dinner. "Not to keep up with the Joneses," for example, and "to release the children when they are of age." I came first and went first—away to college and back after that only as a visitor. Asthma and hay fever were reasons. The dustiest work I could not manage, and harvests were the worst. I never questioned my incapacity and may have welcomed it: it was out of my hands. For farming, I was not fit—4F.

George was not so afflicted and so he caught the farm. He followed me away to college but within a year returned closer to home. A wife-to-be played a part, as did conflict with his desires, which were not settled for him by a fiat of poor health. More outdoorsman than I, he spent much of his freshman year refurbishing a canoe and exploring a river. So college began with a detour. Soon he married, took his bachelor's degree, and a Master of Science, and considered working for agri-business. Then the chance to farm at home arose.

Did George feel released? Was he called back? Did he have a choice? For years I thought farming was what he wanted and that he graduated to his desire. Then for years I guessed that duty called and he was drafted. Neither one way nor the other wholly, but he took up the farm.

Never again did I plow in December following a late corn harvest. To plow after harvest, although it exposed soil to the wind and thus to erosion, meant quicker, more efficient planting the following spring. Corn planted early caught more of the essential spring rains. But the bottoms are windswept and bitter. A canvas sheath fitted to the tractor and rising up around the steering wheel protected me as I aimed north into the wind. Motor heat, mixed with gas fumes, collected within the sheath and wafted back to warm my face and chest. Snow flurries dusted the field. The tractor inched ahead numbingly slowly, like a caterpillar crossing a sidewalk. Under gray clouds, white air swirled. Even then, appealingly, the plow could erase the snow and make a dark stripe across the field, as if, late in the day, I'd taken a wet cloth to a blackboard. But coming back against that stripe, the dark earth was already whitening as I cut the next furrow.

On the return, I was headed south. The north wind blew the motor's warmth ahead of and away from me and wormed into my unprotected spine no matter how many layers I wore. It was as if metal fingers had escaped from a freezer to slither up and down my backbone. I felt peeled and exposed like the tiny, green tongue of a new winter onion. Nor could I imagine, on that slow, southbound crawl, a new protective bulb forming around me.

George found his ways, however, in part by attending to his surroundings—to onions, for example, and to who had planted them, and then asking himself, What makes chrysanthemums canonical?

Springs

In 1959, the citizens of Grand Pass restored their spring. Erosion running off the bluffs had buried it, so they dug it out and lay in tile to redefine its source. A newspaper story tells of their digging through twelve feet of sediment and finding a walnut log at the old base of the spring, a log "on which Indians had knelt to take their drinks." In settlement days, Dover Lake lay just off to the north, a short walk from where I had paused, forgot to drink, but copied down the verse that promised my return. A century before, a wooden bridge that led to the lake had often been lined with persons fishing, bathing, and enjoying themselves. If a farmer needed help, he knew he could go to the spring to hire a man for the day. In the old days, they called it "Shaky Spring," because of quicksand around it before levees, drainage ditches, and clearing turned wetlands into farms.

One story tells of a team of oxen, driven by a slave boy and hauling a wagonload of rails, all swallowed by the spring. The story says "boy," though he may not have been. Or perhaps a boy, black or white, would be more likely to get himself and a wagon of rails into trouble. Another well-known spring, also at the edge of the bottoms, beneath the bluff site of the Missouri Indians, is called Bottomless Spring and identified with the following story, as told by the late Sam Irvine. Clearly it is a variant of the first.

> This fellow had cut himself a load of hay and was hauling it to his barn with a yoke of oxen who had been out in the field all day and were plenty thirsty and, as they came over the top of the bluff, they smelled the water in the spring and started down the hill for it at a dead run. Now you didn't drive oxen with reins so you could haul them around, but you hollered "Gee" or "Haw" or "Whoa." This fellow kept hollering "Whoa," but the oxen just kept on running down the hill to the spring. When they got to the spring the wagon kept on rolling and pushed them into the water; and the oxen and the wagon and the load of hay and the farmer went on in too, and they went down and down and have never been seen again to this day.

One wonders whether the men telling these stories believe them or tell them to see whether you will. Irvine figured Bottomless Spring couldn't really be bottomless because if it were, the water should be hot. So he bought up all the fifteen-pound-test fishing line in the county—fifty spools of it, each spool holding a hundred yards of line—and took a rowboat out to the middle of the spring. There he tied an eight-ounce lead sinker on the end of the first line and dropped it in. It sank right in, like bad news. As that spool ran out, he tied on another line, then another. After a while the line got so heavy he had to hitch it around his oarlock to keep it from cutting into his hand, but the line kept descending into the pool. He got down to the next to the last spool before the weight of accumulating line broke the string, so Irvine never did plumb the spring's depth.

It would require skillful packing to load a rowboat with fifty spools, each holding a hundred yards of line, and leave room to man the oars. I'll assume Irvine was competent at that. By the time the drag of the line caused him to hitch it around his oarlock, though, somewhat more than fifteen pounds was weighing it down. And it's a bit hard to imagine him launching a rowboat on that spring, which looks like a good-sized puddle whenever I've walked around it. Besides over thirteen hundred yards of line is much the better part of a mile. But let's not sink his story. Long before I first saw the spring in the early 1950s, the state had put a protective fence around it. If you climbed over, you were on your own. I have always stayed on the safe side.

In the bluffs above Bottomless Spring is the opening to Madoc's Cave. My father liked to report that a man seeking to fathom it rolled a cannonball in from its mouth, and the ball went bonk, bonk, bonk, as if down steps, and bonk, bonk, bonk some more, with no end to the bonking until it faded from his hearing. I've searched for the cave and never found it. But, then, I know better than to search hard. Madoc and the white, Welsh Indians are another story.

We may not have our heights, but we have depths. The bluffs are steep enough for a loaded wagon to goad oxen into a wild run downhill. The soil is deep, and the springs and caves plunge down and down. Inlanders feel something of the pull of the sea in all this. Standing as farmers do, watching the hay come round or the corn grow, we stare as men and women will, toward beginnings off some half-forgotten shore.

So an old story about finding farmland has a man just off a boat instructed to put an oar upon his shoulder and walk inland until someone asks why he carries a winnow fan. There he'll find good land to farm. That's a turn on the *Odyssey*, or the *Odyssey* on it. Who knows which came first? Odysseus had to shoulder his oar similarly, not to locate a farm but as ritual cleansing after having slaughtered Penelope's suitors. For all his eminence, Odysseus had to humble himself and walk far enough inland to be mistaken for a farmer. Then he made sacrifices before returning to his coastal home.

I wonder about the farm boy who, coming upon Odysseus' oar, saw it was no winnow fan. Lifting it from where the hero had left it in a field, he noticed the scar of oarlocks on the handle. Drawing it across a fence, he found no point of balance as the blade, like a witching rod, dropped down. Pulling lightly on the handle, he steadied it as he felt the stroke begin. Would he have put the oar on his shoulder and walked downland with it until he discovered what he carried?

That is the path eminence has mostly taken. But if everyone took it, who would be left to clean out and retile our springs?

II
Hanging Mart Rider

Dark Cloud

The Hamilton side of my family came from Callaway County, Missouri, which, like Saline County, belonged to an area known as Little Dixie. They had come from Pennsylvania by way of Texas. A cousin of my great-grandfather's fought for the Union and was wounded at Gettysburg. He became superintendent of Gettysburg National Cemetery. My great-grandfather, our first George, is said to have been outspokenly Northern and to have run off with Union troops at fifteen as a drummer boy. His father brought him home then sent him to school in Kentucky. He married a Kentucky woman whose family came from the other side of the fence and had owned slaves. The young couple tried sheep ranching in Texas before settling into farming, sheepherding, lay preaching, and novel writing in Callaway County, which had seceded from the Union on its own, although that lasted only until the first Federal militia crossed its border. But that was enough to redefine the region as the "Kingdom of Callaway," and my father and uncle grew up imbibing the Lost Cause romance of the "boys in gray," my uncle clinging to it longer. Once I heard him qualify like a lawyer the oath required of Southerners. It was not an oath of allegiance, he claimed, it was only a solemn promise not to fight the Union anymore.

Another time, Unc and I were riding along in his pickup, going from somewhere to somewhere else in the bottoms, and in the midst of another discourse, Unc paused and said, "Dark cloud goin' south." I glanced at the sky, saw nothing, and glanced at him, puzzled. Unc grinned and pointed with the stem of his pipe. There on the roadside, on foot, was a black family, two adults and four kids, walking our way, south. We sped on by.

Before the Civil War, in Saline County as in many other places, blacks needed passes to go off a master's land, and "patrollers flogged" those caught without them. Before emancipation, too, "Free Negroes" lived "subject to good behavior" and often under a bond. Bonds ranged from one to five hundred dollars. As part of a bargain for statehood, the territorial legislature promised that an "antifreedman clause" in its constitution would not be enforced. But free blacks had to come up with their bond then demonstrate their ability to achieve self-support "by legal labor," which, in practice, all but bound them to slave owners.

Both before and after the war, a large part of that labor was in hemp. Hemp had a long stalk and was heavy to cut and carry and laborious to strip and form into bales or "breaks," which then had to be wrestled to the wharves. Hemp was the cotton of the region, the crop most dependent on the cheapest labor possible. Accordingly, James Thorp describes Saline as one of the top slave-holding counties in Missouri and lists three of the county's top ten owners as from Miami with between twenty-eight and thirty-three slaves each, though the largest number attributed to any one owner in the county was eighty-eight. Growing hemp on a thousand-acre farm made cheap labor attractive.

The WPA oral history tells of a man who was chained to a hemp break and froze to death. This story leaps into mention more than once, never with any detail. But apparently a master, wishing to reprove some offense, chained his slave to a pile of hemp one winter night, and the man died. Except for the want of a mob, you could call it a lynching.

Another story begins similarly but differs in its outcome:

Mr. Isaac Kile, a prosperous Miami bottom farmer, suspected that his oat bin was leaking oats too rapidly, so to verify his suspicions he set a steel trap at the opening of the bin one evening last week, and going out at bedtime found a man fast by the hand. As the man seemed fond of the trap, Mr. Kile left him with it until morning, when reading him a lecture on the Eighth Commandment, he was released and told to sin no more.

This account from December 1874 suggests another cold and threatening night. Perhaps Mr. Kile was as indifferent as the slave owner to the possible consequences of his trap. The lighter tone of the story, though, suggests this sinner was white.

After emancipation, the state Vagrancy Act permitted the sale of freed slaves back into slavery for terms of six months. So in 1874, Nelson Thomas was sold to Edward Dance for six months for thirty dollars. Dance is identified as, formerly, a prominent slaveholder, though not prominent enough to have made the top ten. Nine years later, Joe Boatright was likewise found guilty of vagrancy and so punished. Bidding started at ten cents and ended at six dollars and fifty cents. The buyer was the prosecuting attorney. One would like to think that was a joke, that the attorney paid his six-fifty and let Boatright go free. There is no such evidence, however, and even had that been the case, there must have been much hilarity at Boatright's expense as the onlookers examined the merchandise and boosted the bidding up over six dollars.

Often unnamed blacks populate small disasters. In April 1877, the *Miami Index* reports that the "Fannie Lewis on her down river trip rammed her barge against a pier of the Boonville bridge, drowning six or seven Negroes, sinking the barge with a loss of 10,500 bushels of corn and six hogsheads of tobacco. Insurance $6000." Its exact report of material loss underscores its indifference to the number, much less the names, of the men who drowned.

More frequently, the tone is that of condescension masked as entertainment, as when an 1878 issue of the *Index* reported,

> A young man from the country on last Tuesday while in town, was so far removed from all restraint from his liberties as to get drunk; while in that state he made an assault on the Negro errand boy at McDaniel Bros., got knocked down by the Negro, and was then fined $5.

That errand "boy" was probably not a boy. Otherwise he comes off pretty well. Perhaps we could venture that a certain, delicate-to-define division of mind begins to flicker in the attention focused on blacks in what had been a slave-holding county, and a slave-holding state, though a state that had remained in the Union. So, from the WPA oral histories of the 1930s comes this notice of "Colored Baptisms":

Quite a number of their white friends walked down to the appointed place, disposing themselves on the slopes of the hill, under the shade of the trees to witness the ceremonial.

Or from another old newspaper:

And now comes the case of the State of Missouri against Larkin Brown, a colored lad of Fairville, in which Mary Brown (col'd) whose fighting weight is about 250 pounds avoirdupois, swears that the said Larkin Brown did, on or about the middle of June, assault and beat her with a stick, and did also with murderous threats thrust a loaded pistol in her face, greatly disturbing her peace and dignity; and furthermore says upon her oath, 'if you don't do somethin' wid dat nigger, I'll kill him sho!' Whereupon, the court fined Larkin $2.50 and costs.

The writer must have enjoyed rolling "av-wha-ar-du-pwaa" on his tongue, the name, Larkin Brown, and the dialect, all of which composed his own peace and dignity; much as, for their own pleasure, white spectators of the baptisms disposed themselves at a safe critical distance from the "ceremonials."

But that entitlement, which will seem mild to some from a distance, cannot mask a harsher reality. An article of January 1, 1911, entitled "Death's Harvest in Miami Township, 1910," lists the thirty-one deaths, "16 white, 15 colored," of the year before. The nearly equal numbers suggest more racial equality, in number anyway, than one would gather from reading the papers or from the one in five ratio of early censuses—except for one detail. Five of the six infants who died were black, as were two of three under twenty. Eight of the nine who lived to sixty-five and beyond were white.

Or again, from the *Miami Weekly News*, April 8, 1886:

Jeff Wilson col'd was hung for murder at Lexington, last Friday, in the presence of about 5,000 people.

The prayer delivered by Rev. Dr. Howard, now conducting a revival at the Col'd Baptist Church in this city, was a splendid effort.

Coon Hollow

Once I found a geographical poem devised as a mnemonic for just about every locality in Saline County, but it leaves out Pennytown, Cow Creek, and Coon Hollow. Each was a post–Civil War black settlement. Coon Hollow lies between Miami and the old town site of the Missouri Indians. It fringes the drainage of Coon Creek from the Tetesaw Plain into the Tetesaw Bottom. We drove through it again and again, and when I was away at college, I highlighted my rural background by identifying myself as from Coon Hollow. In a room dotted with graduates of Andover and Deerfield, I could count on that for a laugh. For my senior yearbook, I identified myself again as from Coon Hollow, but the shadows of the early sixties must have dimmed my so doing. I had just become aware of the Congress of Racial Equality and had been drawn toward, but remained shy of joining, its few tiny demonstrations. I delayed and delayed picking up my copy of the yearbook and only did so the day before I left college.

Freed slaves were granted land in Coon Hollow, though few of them kept theirs long. In 1915, black children in Miami had a separate school, two teachers, and a new building, which the parents built themselves, the district having bought the land. Rather than a two-story, Georgian structure with a Latin motto and busts of Shakespeare and Jefferson by the door, it was a small, clapboard, one-room schoolhouse like those that dotted the bottoms. Like the other one-room schools, it went only through the eighth grade, which was as far as many a farm boy managed. The community overall felt that plenty good enough for blacks. Coon Hollow had such a school. Every little town, though, had a high school for white students able to go on.

Since the bottoms flooded regularly and no house stood on our new farm, we found a house in Marshall. My father and uncle commuted to the farm. In town, by an exception that I could never reason through, in an era of segregation of both neighborhoods and schools, a black mother and daughter lived two doors west of where we first lived on Eastwood. Early on, testing the taboo language of new friends half under my breath, I called the daughter "nigger" as she passed me on the sidewalk. I was seven, crouched low, sitting on the curb by a playmate, and I knew better. The year before in an integrated school in Illinois, I'd held hands with a black girl during first grade recess games and we had smiled at each other.

Peggy was an adult. Dropping the coffee pot she was carrying, she chased me across the street and around two houses before I escaped into a deep backyard. She might have given me a whipping had she caught me, and my parents would have approved. Instead, she made her complaint to them, and I made a trip that night to apologize to Peggy at her door. We were always cordial after that but never intimate. Only late in high school did I sit and chat with Peggy and her mother on their porch, for there was always a levee between us that was easier to maintain than to level.

Reverend Harvey Baker Smith lived only a few doors away on Eastwood. Reverend Smith wore a cutaway when he preached, and his family came from Virginia. For a time, I led the youth group of his church, and one summer I got into the habit of going around to his house to watch the Friday night fights. It was the era of Rocky Marciano, Jersey Joe Walcott, and Ezzard Charles; and it was disturbing, Reverend Smith said, and wrong, to see a black man and a white man fight. He said "black man," which sounded strange to me, the slogans of the sixties being more than a decade off. I caught its accuracy but not its connotations. He looked worried, as if he wrestled with the Almighty, while he stared hard at the fight and let his ice tea, served by his grown daughter, dampen the magazine on the floor beside his chair.

All through my school years, I walked the north side of Eastwood. Conway joined Eastwood about halfway along its one mile length. Tarred rather than paved, it extended a couple hundred yards north, hooked around to the east, and dead-ended. A cluster of black families lived on it. My friends and I called them Negroes. We knew few faces and fewer names. On most mornings, several black kids would appear from Conway as we were making our way to school. They would cross Eastwood and walk up the south side, in separate procession, as we made our own way on the north. A block before we came to our grade school, they would turn south again and walk several blocks more to their school, Lincoln. We never spoke to each other as I recall, hardly even eyed each other. "Dark cloud moving west, then south." So that group appeared to us, while our own dim cloud paralleled them for the long block we shared.

As I grew older, I'd often walk Eastwood alone, going or coming, at dusk or after. All through high school I'd feel an edge of unease if I passed a black boy, or man, in the dark, though usually he'd be on the south side of Eastwood and I on the north. If I saw him looming ahead on my own sidewalk, I might cross over myself, affecting an attitude to suggest my real purpose took me there. When I became embarrassed at that poor tactic and chose to hold my ground, I'd steady myself, bring any humming or whistling to a halt, and control my measured pace until we passed.

We probably looked at each other, probably nodded. Often I'd break into a run a few steps later, reasoning that I often ran my Eastwood mile anyway, for practice. If I saw the stranger from far enough off to make it seem I wasn't running just because of him, I'd pass him flying, holding my stride steady, one step for each of the five-foot sections of the walk until I was out of sight. My shadow pursued me, shortening as I ran under a streetlight, then leapt ahead, lengthening as I passed it.

My high school days began with the Brown vs. Board of Education decision. Before that, in Marshall, the county seat town, blacks went to Lincoln School, which, like the Miami school, went only through the eighth grade. Students who aspired to more education took a bus daily to and from Sedalia, thirty miles farther south. So much for "separate but equal."

While in the eighth grade we learned of the Brown decision and that blacks would join us the following autumn. Late that spring, a teacher protested that "it just wasn't right. The coloreds and the whites shouldn't mix. It would lead to the mongrelization of the species. God had made the races separate and so He had meant for them to stay." She "wouldn't stand for it." Even we recognized a fallacy in that last step. Carolie, the girl next door of my youth, put her hand up quick. "Maybe God has changed His mind," she offered.

The next year, black kids filtered in among us, and we learned some names. Walter and Eugene, for example, came off Conway to join our classes. But when we were on Eastwood together, as long as I can remember, they walked the south side and we the north.

I remember one exception. Carolie, Eugene, and I were walking home one afternoon during our junior year of high school. Clearly this was an experiment. But overtaking us by car, Carolie's mother squelched it. "You get in here right this minute." Her invitation was not general. Carolie reminded me of that story at a reunion after I'd reminded her of her response in our eighth-grade class. Both of us were telling our stories to Eugene, testing ourselves, and he was listening and testing too, letting us hear him say "black."

Eugene had been among the first handful of blacks who had joined us that first autumn of integration, six young men who also joined our football team. Each one proved himself a player. We didn't try this story on each other, but Eugene probably remembered an early team meeting and our coach at the blackboard, diagraming a play. He had his Xs and Os on the board, and lines ran from one to another, indicating an off-tackle slant with all the required blocks. If everyone did his job, the runner would have only one man to beat, far from the point of attack, and he should score. But the lines weren't joining up; something was amiss, and our coach stood at the board scratching his head through his curly red hair. After a moment he murmured, more to himself than to us, "There's a nigger in the woodpile somewhere."

The room had to have felt our collective intake of breath. One boy giggled but slammed the door on his tongue. The coach's face turned the shade of his hair; he stammered an apology and went back to his secondary struggle of diagraming the play. He made no second slip.

In an odd parody of integration in that first year, our junior varsity football team, but not our varsity, played the black high school in Sedalia. One Wednesday night we rode the bus to Sedalia and played under streetlights—a strange homecoming for several of our players. No bleachers, no crowd, only a few bystanders from the neighborhood. I started as our quarterback and heard my first trash talk, "We goin'ta smeah yo ass, whiteboy," from a wild-seeming fellow, crouched and grinning across the line of scrimmage. He did several times, most efficiently, and we lost.

Frank James

"I'd know that man's hide in a tanyard."

"Where did you come up with that one, Unc?"

"Like that? Guess it reminds me of a story I heard Sam Irvine tell of two neighborhood boys and old Frank James. Those were mean years, after the War. The James boys had ridden with Quantrill and turned guerrilla; wasn't much left for them round here. Except outlawing. Much the same might have happened to ol' Doc Benson, had they not caught him first. Course it may have just been their natures.

"That was the winter Pinkerton's men tried to get the Jameses by bombing their house. Late January 1875, a Friday. Five or six of Pinkerton's men snuck up and tossed a bomb in the window. Bomb went off all right; only trouble was, Jesse and Frank weren't home. It killed a half brother of theirs, just a boy, nine or eight. And it maimed their mother. Tore off half her right arm. So later that winter they brought her over here to mend, to a farm out on the Tetesaw Plain, just south of the old Missouri settlement. They left their mother in a cabin with friends and went on about their business. Then later in high summer, Frank and Jesse rode back.

"There was still a lot of prairie back then; it wasn't all corn and beans. They say a man sitting on horseback could tie the big bluestem up over his head. These plains were a friendly place for the Jameses. There's a little grove of trees out there, some four or five acres, and they'd rendezvous in it when on the run. From there they could see riders way off in any direction and make a break for it, down into the bottoms, into the deep timber.

"A man from Texas had brought up sheep and longhorns and was grazing these plains, and he'd told some neighbor boys, a couple of brothers, they could salvage the wool from any sheep they found dead. Those boys got their spending money that way. Your daddy and I did about the same when we were boys, running traplines over in Callaway County. Skunks fetched the best price; all the boys did it. You should have smelled our school rooms some mornings. What our teachers put up with.

"Sheep can be a mean business, specially during a hot, dry summer. Your great-granddaddy herded sheep down in Texas, then over in Callaway. Some years lots of lambs don't make it; some sheep don't either. Those boys kept an eye out for buzzards then headed out on the plains to beat 'em to it. After a sheep has been dead, don't have to be too long, you can pick the wool right off, like skinnin' a rabbit.

"You can imagine how it was, in the deep, still grass, under hot sun, flies and mosquitoes buzzing around, and the no-see-ums. Little clouds of aggravation round your head, not to mention dust and pollen. Buzzards overhead, giving you an idea of where to look. But it could be pretty, too, with the sunflowers and coneflowers, both purple and yellow, grinning.

"Well one day these boys came upon a dead sheep but didn't have their wool bag with them, and one ran home to fetch the bag while the other waited to keep the buzzards off. He lay down not too far away, giving those fellows something new to contemplate. About then the James boys came riding over a hill, maybe that one, on their way to visit their mother. They must have seen the buzzards, circling and circling, and so came riding over to investigate. Frank saw the boy down there in the grass, but he saw something else, too, since he sat high in the saddle. So he trotted on over and without any ceremony demanded, 'Get up behind me, son.'

"'I've got sheep to wool, sir,' the boy replied, he'd begun picking at his work already. Scraps of wool stuck to his fingers and to his shirt.

"'Get up here quick,' Frank insisted, 'move now, right smart.'

"'Go on, Mister, I know my way home. I've got sheep to skin 'fore it gets dark.'

"Now Frank, was getting impatient, and he more or less hollered, 'Jump, boy!' and that time that young fellow jumped right up behind Frank to see how the tall grass hid free-ranging longhorns, a whole arc of them, bearing down exactly on where he'd been lost in his work. Guess he lost the wool, but the James brothers got him home safe and sound and went on their way.

"Well, those boys grew up and stayed around here doing one thing and another. Frank and Jesse made their doomed raid on Northfield and were all the more on the run till Jesse was murdered and Frank turned himself in. He was acquitted, you know. Two juries let him off. Had lots of sympathizers. After that, he kind of eeked out a life, trading on old notoriety. One thing was he'd travel round to the fairs and start horse races. Guess the gun that he carried still carried some weight.

"However that may have been, they say that Frank never forgot a face. The boy he'd picked up heard about that and lay in wait for him. Coming into Marshall for the county fair, Frank would stay in a hotel just off the square. And the boy, a grown man then, parked himself on one of those benches and waited.

"And waited. What a way of wasting a farmer's morning, Frank not being what you'd call an early riser. But sure enough, long about midmorning, Frank came strolling out of the hotel to find a grown man intercepting him who, with some slight ceremony, asked, 'Morning, Mr. James. Do you remember me?'

"Frank glanced over him quick, and who knows what he saw? Maybe a trace of dust from these loess hills. Maybe the shadow of clouds and coneflowers. Or maybe a whiff of dead sheep and the sweep of a buzzard's wing. Whatever caused it, Frank's smile swept over him, like a buzzard's shadow on grass. 'Still in the sheep business?' he asked as he passed."

Which is a pretty story for a country that still favors individuals and individualism. We take Frank James as a distinctive man and the boy, when grown, to be his own man too. We understand the tale in these terms without much of a second thought. But there could have been other means of recognition—as Southern sympathizers, for example, or as small-account farmers, as brethren who, without much scratching, could each see beneath the veneer of the other. Sam Irvine gave the boys' name as Tucker. Several Tuckers are listed in Confederate companies raised from Saline County during "the War," and several Irvines rode with them.

Recognition was ever the problem, and the wrong sort could, so to speak, put one's hide in a tanyard.

Dr. John Benson

A correspondent of my father's, a native of Miami, wrote from Kansas City soon after Father had become Miami's mayor:

> In my mind's eye I can still see old Lee Kinney, one-legged, black darkey, sitting in the shade of the frame building that used to sit where the Miami Cash Market now stands. He is sitting on a wooden-slat bread box; the chunk of wood with the old rubber shoe heel nailed to the bottom of it takes the place of one leg that he lost between a hawser and a cavil on a steamboat. He plays his french-harp, and soon a crowd of young boys is gathered around him. He has his own nicknames for all of them, Hanky for me.

Local color and small town history always merge. The writer tips my father off to James Thorp and his scrapbooks. Thorp began his scrapbooks in 1949 and compiled a careful list of all Miami High School graduates, from the founding of the public school in 1876 to its consolidation with larger schools in 1948. Thorp is not among

them. Since he is listed on a primary school honor roll (for atten-
dance) in 1897, his graduation should have taken place within ten or
eleven more years. But in that era many boys did not make it that far.

Thorp never married. He worked on a snagboat, winching out of
the channel large limbs and whole trees that the river swept along,
endangering traffic. Paul Stonner, a Miami neighbor, remembers that
Thorp "knew a little civil service," which means he could read and
write pretty well, and he "schooled local boys" who wanted to take
the test. He was also a mail carrier. He bought a Model A and put
taller Buick wheels on it so he could get through the mud without
dragging. Those were narrow, old-time tires, taking fifty to sixty
pounds of pressure, not the new balloon tires that bogged down
quicker in the mud. Then he'd wait until spring for his vacation. In
the spring, when the roads got really bad, Stonner's brother would
carry mail for Thorp on horseback. Once that Stonner boy took four
tires out to some farm, draped over the horse's neck, parcel post.
Thorp must have been about sixty and retired when he began "to
build a book," as Stonner put it.

My father's correspondent also mentions Mart Rider, a locally
notorious villain hanged and buried in Miami. "As boys we used
to get a thrill out of looking at his tombstone." And he brings up
Dr. John Benson and his execution during the Civil War. He wants
to know Benson's exact dates, which should be "one of the easiest
things to get," for his tombstone is in the town cemetery. Alas, however, "I have let it go every time I was in Miami. . . . I would like
very much to have those dates, as some of the fancy stories regarding Benson depend for veracity on the date of his death." In the margin, my father has penciled in, "John W. Benson, Born 3/23/36, Died
10/6/63," for evidently he took the hint and looked.

No subject dominates Saline County memory like the Civil War.
Thorp called it the "Un-Civil War," and nearly everyone still recog-
nized it as the War late in a century that saw two world wars. The
Civil War, however, ranged over home country.

Battle came quickly to the frontier. Missouri's Governor Jackson, a
Saline County man governing a slave-holding, southern-sympathiz-
ing state and fully in accord with those sympathies, cached powder
throughout the state and especially on farms along the river, including several in his home county. He planned to keep that powder out

of Federal hands. One of the first land battles of the war was a Union rout of ill-equipped and ill-organized Southern forces along the river just east of the county as the Union army tried to halt Jackson's sabotage. But Southerners jumped into action, and Wilson's Creek and Lexington were sites of Southern triumph in Missouri during 1861, the first year of the war. By the next year, Federal forces had gained control of the state, although guerrillas rose up to challenge the militia.

Benson was one of those guerrillas. He rode with Quantrill and took part in Quantrill's attack on Lawrence, Kansas. That occurred August 21, 1863, only a month and a half before his execution, which has everything to do with those "fancy stories." Forty-six days make them less probable. Stories tell, though, of Benson's being imprisoned for some time, and of the enduring reluctance of a local militia commander to bring him to justice. He let proceedings drag and deferred the execution once the sentence had been given. The officer is even said to have encouraged Benson to escape. On the way to the execution grounds, he may have let Benson know that should he make a break for it, the attending troops would be sluggish. But Benson made no move; he rode on calmly to his fate and was shot sitting on his coffin. Three firing squads had to be summoned before willing executioners could be found. Requesting only that he be "shot below the face," he "bravely and cheerfully received his death." When his executioners examined the body, they found a lock of golden hair where the bullet had pierced his heart.

Dr. John Benson was twenty-seven when he died, a young man who might well have inspired romance, romance being most of what the South had to offer. That first land battle took place about a day's trip downriver from Miami, with Confederate forces scattered and on the run pretty much with the first cannon volley of the Federals. Having no cannon themselves and unable to get within rifle range, the Rebels broke for home, their pride and equipment in tatters. At Lexington, the Southerners cornered a Northern force they outnumbered four or five to one, laid siege, moved forward under soaked bales and hemp breaks, and forced a Northern surrender. But another Southern troop of 750 new recruits was surrounded and nine-tenths of them captured before a remnant made it out of the county. Caught by surprise, his troops surrounded, the Southern commander ran up a white flag on Blackwater Creek and asked for "time out" of a sort. He wanted half an hour to consult with his staff. "Half hour's too long," the Northern captain replied, "it's getting dark." The Southerners parlayed quickly, surrendered, and sat out the rest of the war in a prison in Alton, Illinois.

The Battle of Marshall, October 13, 1863, came midway through the war, only six weeks after the raid on Lawrence, and exemplifies much of the war in the region. A Col. J. O. Shelby, popularly known as Jo, had raised a force for the South, with many of his men coming from farms between Grand Pass and Miami, that is, from the Tetesaw Plain and Bottom. These men had seen action in Arkansas and along the Mississippi when Shelby decided to raid his home territory and summon more recruits by his victories. Moving in from the southeast, Shelby's force approached Marshall, which garrisoned a Northern force commanded by a Gen. E. B. Brown.

Each side had between a thousand and fifteen hundred men, and both thought the other side larger. Brown had heard frightening reports of Shelby. Shelby assumed Brown's men had been joined by another division stationed one county to the south and so was twice the strength of theirs. The terrain on which they met was divided by Salt Fork Creek and by numerous ridges and gullies. Later it became part of Marshall's City Park and the site of numerous capture-the-flag contests when I was a Boy Scout. Oak, scrub oak, and hickory obscured broad views; neither commander could get an accurate idea of the number or disposition of his enemy. Shelby's army spread out along the creek and the ridges to the east. They approached what they assumed was a defensive line of the Federals with plans to break through. If they could, their way was clear to Miami and the northwest portion of the county, to the bottoms they knew well, to the safety of home.

The day turned into one long episode of hide-and-skirmish. Later accounts, especially ones written to glorify Shelby, describe bold charges with heroes wading in blood. The nearer truth is a day drawn out, tiresome, and dangerous. Many shots were taken, but few soldiers fell, although most everyone felt besieged and spent the day anxiously. The likeliest tally is one Union casualty and five Confederate. Another half dozen Southerners died of their wounds soon after. The Northern army used muskets, and their larger caliber balls did more damage than those from the carbines and pistols of the Confederates. During one pitched battle, a cow strayed between the lines, and her owner, a man named Mitchell, walked out "amid the storm of whistling bullets and screaming shells" and drove her back to safety.

From our remove, it seems impossible to know how hapless an action this was. "Hapless" is the most tempting reading. Nevertheless the Northern army cut Shelby's line and drove off about half of it. Those men retreated to the south rather than toward land they knew better, and it took them another week to link up again with Shelby. Shelby, meanwhile, thought himself trapped in ravines north of Marshall. He advised his remaining men that they could surrender if they wished, but since many had been captured previously and then released only after taking an oath of allegiance that they had now, quite clearly, broken, they awaited court-martial and summary execution. He intended to fight, which he did, through a very thin line posted by Brown, after which Shelby's men straggled through a long retreat across Muddy Creek, the Tetesaw Plain, and into the relative safety of the bottoms. They were enough on the run that they managed few and brief exchanges with friends and family. What glory they found was hardscrabble at best and exhausting. Meanwhile, Marshall remained in the hands of Federal forces who had just one week before executed Dr. John Benson.

Some say Benson rode recklessly in battle, even seeming to rush headlong after death as oxen will rush downhill toward a spring. In the raid on Lawrence, he turned back when informed of a young girl who was ill in a house they had passed. Benson stopped and cared for the girl and took a lock of her hair as a keepsake and wore it over his heart—so it is said.

At the start of the war, people cast about, uncertain of how or where to be, and often moved where they were able. At that time, Benson was engaged to a young woman from Miami. She and her family moved to St. Louis, or somewhere farther east, to be less in danger. St. Louis was under Federal authority. But before leaving Miami she encouraged Benson to shelter a friend of hers who was on the move in much the same way from farther west. So this second young woman came into Benson's home, when his fiancée had already departed from Miami, and his attention to her was not precisely what true love required.

When word reached her, Benson's fiancée turned cold, like winter winds over the Tetesaw Plain, the harshness of which makes Fairville, at the edge of that plain, seem misnamed. She suggested that he might at least prove he knew his duty, which was to the Cause, which led to bushwhacking, Quantrill, and Lawrence. The lock found over his heart was from the younger girl he treated, who reminded him, it is said, of "innocence and truth."

While not impossible, this does seem "a stretcher," as another Missouri river rat would have said. Order Number 11, forcing evacuation of western Missouri counties so as to deprive guerrillas of local support, was issued after the Lawrence raid and because of it. Historians have called it the second most draconian measure ever taken against one's own citizens in the history of the Republic (the first being the World War II incarceration of Japanese-Americans). The second woman's coming to Miami would most likely have followed from rather than preceded Order Number 11. Then, too, the reputed reluctance of the local militia commander to order Benson shot, or of men to do the shooting, does not square with most stories of the day.

But how else do you build a book except by including all the better stories you find?

A Man

A man of Southern sympathy came into the county and tried to lie low, living quietly. Two other men, who had first signed up with Confederate troops but then recast their allegiance, denounced him as a harborer of guerrillas. A detachment of Federals rode up, called this man to his door, and when he tried to escape out the back, gunned him down for resisting arrest.

Another man, a friend of the first, then joined the guerrillas and rode back into the county with men who had ridden with Quantrill. They arrested the two fellows who had denounced that quiet Southern sympathizer. Off across county they rode on horseback, stopping at a distant farm for dinner where everyone seemed "in the highest good humor." Then they all rode on into the bottoms, where the two arrested were swung off their horses, their necks noosed to convenient limbs, and their bodies taken full advantage of for target practice. "Since the war, their skeletons were found, conveyed to Miami, and buried by the citizens."

The first man's friend, the chief arrester of the other two, was killed himself near the county line a few months later.

A man who had participated in an early victory on the Southern side returned home after the capture of his company. There he attempted to conceal himself to avoid taking the oath of allegiance. Denounced, however, as one who harbored and aided guerrillas, he was visited by two men of the county militia. They appeared at his door, requested that he acquaint them with a road through some fields, and then shot him when he bent from his saddle to open a gate in their path.

A Union man who had been around Miami for several months was shot and killed by Confederate guerrillas in that town. They claimed he was a spy. He was certainly not a soldier, had no evident business, and had been seen, they said, in the company of Federals but a short time before. Chasing him to the riverbank to the west of High Street, the guerrillas shot and killed him at the water's edge.

A man of Miami township, a Union soldier, was captured by guerrillas and held in the Tetesaw Bottom. When he asked for water, two guards escorted him to the bank, where he knelt, drank, and was shot and rolled into the river.

A man who served with Shelby was left behind wounded after an action in this county; he took refuge in the brush, dug a cave into the ground, and hid there, until he became ill and required a doctor. The doctor refused to prescribe without informing the militia, who captured the man and paroled him but kept watch over his recovery, after which they arrested and shot him. His mother went to the commander and begged that he spare her son; then she asked to witness his execution. The commander denied both requests.

A colonel and two soldiers on a recruiting mission for the Confederates were shot while sleeping in the timber north of the river. Confederate troops soon surrounded the nearest town and arrested four Federal soldiers and three citizens. They brought all seven to

Grand Pass. One of the soldiers wore the boots of the colonel. Tried by an impromptu court-martial, all seven were found guilty, lined up, shot, and left in a ravine just east of the spur of the Santa Fe Trail that runs down to Everlasting Spring.

Four guerrillas tied their horses to trees and ventured into a friend's house for dinner. While they were within, a Federal company attacked. Though three were wounded, all four made their escape. Two concealed themselves in a house nearby, but during an ensuing charge it collapsed, killing one of them. The other was taken to town and executed while dying from his wounds. A firing squad propped him against a fence and acted quickly. The other two guerrillas escaped, and one swore he would murder the blacks who had informed on his brother, himself, and their friends. After the war, he returned to the house of that fatal dinner and killed an old, harmless black man and wounded another. The men who had reported on them had long since departed.

A Union man who had lived for ten years in the Tetesaw Bottom was accused of being an informer and shot. He lived for over a week with his wounds, and he named his five assailants. It was claimed that his death was in retaliation for his having reported the whereabouts of that dinner party.

Late in the war a man, a Whig who favored neither abolition nor secession, was living above the upper bottom. A Federal captain sent two men to his house for dinner. Later the captain sent more men disguised as bushwhackers who asked for information about the soldiers lately hosted. Though warned that the newcomers were probably militia, the man assumed he was among friends and pointed out the path of his recent guests. The militia asked that he accompany them a little on their way and then shot him three times in the head and twice as often elsewhere as soon as they were out of sight of his house.

Sixteen bushwhackers came to a blacksmith south of Miami, commandeered his forge, and set about shoeing their horses. Meanwhile Federal forces, both cavalry and infantry, were led toward the place; but it was raining, muck prevented coordination of their attack, and the bushwhackers escaped. Thereupon the cavalry prepared to burn down the blacksmith shop and would have done so had not their guide assured them that the smith was a Union man.

Two dozen militiamen were surrounded and attacked by about a hundred guerrillas led by one of Quantrill's men. The Federals, however, had possession of a brick building and held their attackers off. Under cover of darkness, they retreated, abandoning their horses. Meanwhile, a number of the guerrillas had been wounded, including their captain. These men were left in cornfields nearby, where the captain received daily ministrations from a young woman of the neighborhood. Later the militia returned and arrested the owner of the farm, and it would have gone hard with him had the woman not turned herself in as the nurse of the guerrillas. Within a few weeks, she was released.

One night, a man, a judge in Miami, heard someone banging on his door. Fearing bushwhackers, for he was a Union veteran, he answered nevertheless, there being no backdoor escape. He found two young persons who demanded that he marry them, which he did without even making them wait while he finished dressing.

A Southern captain, released after giving his oath, journeyed upstream from Vicksburg to St. Louis to Miami. Well aware of conditions at home, he considered going right on to Montana, where the steamboat captain offered to take him. At Miami, however, he risked the worst and disembarked. Presenting himself to an officer of the Wisconsin Cavalry stationed in Miami, he asked for protection. The officer assured him of that so long as his company remained in the area; however, he expected new orders. The captain went on home, those orders came, and the cavalry headed south. Soon the militia came to the captain's house and took up positions on three sides so as to encourage his flight to the rear. Instead, he answered them, unarmed, at his front door and asked what they wanted. "Water," they replied, and after receiving it, they departed.

A company of guerrillas, fresh from successful skirmish, came upon a Union paymaster's escort on the Tetesaw Plain. The escort had paused to feed their horses when the guerrillas fell on them. The soldiers raced over the bluff and down into the timber in the bottoms, leaving behind the clerk in his wagon. The guerrillas mistook the escaping escort for the greater prize and pursued it while the clerk threw a small iron safe containing $250,000 in greenbacks into the tall grass. When they turned around, the guerrillas took the clerk's watch and rode off. The Union recovered its money.

A squad of soldiers dressed as civilians and went one evening to the home of a quiet old man who lived east of town and had taken the oath of allegiance. Announcing themselves as bushwhackers, they requested food, horses, and a place to hide. The man claimed that he had no more than a pair of socks to offer, which the soldiers took. Then they said, "You have sworn to report us." "I will not report you," the man replied. The soldiers returned to town and waited until the middle of the following day, but when the quiet old man failed to make his report, they rode out again, escorted him to a nearby ravine, and shot him.

Judge Lynch

In 1824, a Saline County slave named Jack charged his owners with assault and battery and false imprisonment. Suing as a pauper, Jack found rapid support as the judge warned the defendants to "permit Jack to have reasonable liberty of attending his counsel . . . , that he not be taken or removed out of the jurisdiction of this court, or subjected to any severity by reason of [his] application . . . to sue for his freedom." Within the year, the owners had freed Jack and paid all costs, and Jack had dismissed his suit, having won the understanding that were he still a slave, he'd have rights his owners "were bound to respect."

It is striking that Judge David Todd allowed this case, much less guided it to this decision, thirty-three years before the Supreme Court's Dred Scott decision, which was of a decidedly different character. It provides sufficient evidence for our human capacity for devolution being at least as great as the reverse, since by midcentury terrorist tactics had become commonplace against blacks.

Saline County was prime agricultural land, which invited large farms and the use of slaves. Almost an island, it and the rest of Little Dixie stood all but surrounded by free territory. Kansas, Nebraska, Iowa, and Illinois, contiguous states to the west, north, and east, were all free. And it was cut off from the South by the Ozarks, where smaller, hardscrabble farms were less conducive to slavery even if their owners were no more pro-black. Most Little Dixie settlers were from Kentucky, Tennessee, and Virginia and had brought slaves with them.

Each slave was expected to work around ten acres of hemp, or half as much corn, and their individual values were in the hundreds of dollars, whereas the cost of unimproved land was well under ten dollars an acre. As the region filled with farmers, slaves were assumed to provide the competitive edge even though by early 1860 prime field hands—young, strong males—were worth as much as $1,600.

Meanwhile, there was no snake lower than an abolitionist, as my uncle would likely have said. By midcentury, an almost millennial fervor raged as widespread and increasingly shabby guerrilla warfare began. Missourians moved into Kansas to vote for slavery in Kansas elections and to bully whoever got in their way. Kansans raided Missouri farms to free slaves and burn whatever they could torch. Slaves slipped away through the Underground Railroad or by other means. Many were returned by outraged and outrageous force. Individual blacks were trusted less and less. Patrollers were more than ever "necessary."

A certain conflict of interest arose in the courts. Blacks accused of capital crimes were also property and not readily hanged or lost to a long sentence. An owner, foreseeing trouble, often sold the offending slave quickly so as not to lose his investment. In the general understanding, an alarming number of blacks "escaped justice." A slave belonging to the same Judge Todd who had favored Jack was charged with murder and tried four times, with hung juries each time before it was Todd's turn to ascend the circuit court bench. Todd excused himself from ruling either way, citing his conflict of interest, but he passed the case on to the next county, where a fellow judge dismissed it.

Anyone fraught with crises of conscience found long-established precedents offering ways out. York, for example, had been William Clark's slave and servant on the famous westward expedition. When the expeditionary force reached the mouth of the Columbia River after a year and a half of travel and faced crucial decisions about how to spend their second winter, each man and woman had a vote, including Sacajawea and York. But when they returned to St. Louis, York found himself a slave once more. He asked for his freedom, but Clark, the first governor of the Missouri Territory, "could not part with him," and it took another ten years for him to discover his ability to free a man. By then, very likely, York was worth little.

In 1860, Abraham Lincoln did not win a single vote in Saline County, though a total of 3,698 votes were cast. Four years later, he won, 170 to 98, the vastly reduced numbers due to the oath then required to vote—or to preach, teach, practice law, or to do nearly anything else—which did little to help Southerners reconstruct themselves.

In May 1859, with war in the offing and abolitionist sentiment closing in, a slave named John was accused of murdering Benjamin Hinton. Hinton and a partner had run a small business in the Tetesaw Bottom cutting timber and selling fuel to steamboats. Their camp was alongside what would become Laynesville between Malta Bend and Miami. Each man had a slave; John belonged to Hinton's partner. Hinton's slave came to his cabin one morning and found his master's head bashed in with an ax, brains on the floor, and a trunk rifled. Before long, John was held with ten bloody dollars in his pocket and more in the possession of his wife. John was taken to Marshall, but the sheriff spirited him away to the next county before the worst could develop.

Within days, a slave named Jim was accused of attempting rape on a local woman. Another named Holman was charged with wounding his owner with a knife. A third "Negro boy," unnamed, was said to be caught naked and attempting outrage on a girl of the Arrow Rock neighborhood. He was lynched that night after a "committee of citizens" found him guilty and his owner concurred. Jim, Holman, and John were brought to Marshall for a special session of the circuit court.

The court met on Tuesday, July 19, 1859, in the courthouse in Marshall. Judge Russell Hicks, from out of county, presided. He anticipated trouble but felt he had the clear support of the sheriff, the jailer, and the county attorney. A mob had formed nevertheless and accomplished its will by the end of the lunch hour. After a morning session found John guilty, the court assembled a second jury for Jim. But Hicks declared a noon recess while determining that, for their own safety, the prisoners should remain in place during the break. Privately, he asked the sheriff to return them to jail as soon as the crowd dispersed. The crowd hung around threateningly, however, so Hicks said that he would escort the men to jail himself.

The sheriff led the way through the crowd as it became a mob, and Hicks fell behind. He was a heavy man with a crippled leg and could not keep up. The small procession reached the jail, but the prisoners could not be secured. A James Shackleford harangued the mob, and when Hicks asked the county attorney to see if he could quiet them, he was told that if such men "had taken the matter in hand, it was all over with the prisoners." John, Jim, and Holman were seized and lynched in a ravine a block north of the courthouse.

> John was about 23 years of age, a valuable slave, worth probably $1500; had an intelligent and open countenance, and conversed very freely with all those who indicated a willingness to hear him while he was chained to the stake. . . . We did not hear of his having made his peace with any judge more terrible than "Judge Lynch." In his agony he prayed more to those around him than to the One above him. He screamed and groaned and implored those about him for mercy, calling on those he knew by name. He lived from six to eight minutes from the time the flames wrung the first cry of agony from his lips, the inhalation of the blazing fire suffocating him in a short time. His lips and arms were burnt, a portion of his head and face, and a part of his chest. His body remained, a charred and shapeless mass.

Both Holman and Jim were hanged. Holman "died apparently easy." Jim the mob had also intended to burn. "To our eye," report continues, "his offense was the blackest of the three, but the law does not recognize it as equal to either of the others." Witnessing John's agony though must have made them reconsider their plan, for the mob hanged him instead, and he "died hard."

Shackleford tried to justify mob action. In the *Marshall Democrat* just three days later, he wrote,

> People may call it mob law. Well . . . it was mob law when the laboring men of Boston disguised themselves as Indians and threw the tea overboard. It was mob law when the People of France hurled the Bourbons from the throne, and crushed out the dominion of the priests and established a new order of things. Incapable of discriminating, they waded through oceans of blood of the innocent as well as the guilty, but they saved France. I know no reason why we should not have a little mob law in the State of Missouri, and the County of Saline, when the *occasion imperiously and of necessity* demands it.

Though he owned 160 acres east of Marshall, Shackleford is not thought to have been a slaveholder himself. Within less than a year, he sold his farm and with his wife and five children moved fifty miles west into another county. Judge Hicks resigned the bench, enraged and humiliated that he could not keep peace in his court. He swore never to reenter Saline County. Union troops, when they entered Marshall, jeered at the citizens, saying, "So this is where you burn a man at the stake."

That was not the end of Judge Lynch's career in Saline County however. In September and November, 1872, two black men were lynched, each for the "attempted rape" of a white woman. In 1900, a prisoner who was attempting to escape wounded the sheriff's wife, and he was also hanged, in the same ravine north of the square. "They say the tree died," adds Paul Stonner, who goes on to report that

> In 19 and 25 there was a black boy down in Slater who was awfully well liked, and another boy they didn't much like killed him. So the boys in Slater were told to "do something about it, if they wanted protection." Some men from Slater came up to Marshall and got a telephone pole, or some big pole, and broke into the jail and got the killer out. They took him down to the Glasgow bridge, tied his legs together with bailing wire and his arms down to his sides and threw him in. A fisherman caught him a few days later. Funny thing was, there was that sheriff sitting there, and those men came with a big pole battering ram and "nobody knew anything about it."

Mart Rider

Once, at the close of the war, an army boat showed up in Miami draped in black. Seeing the boat as death itself and fearing retribution, most Miamians fled. But the boat was only draped in mourning for Lincoln. The captain had stopped so that the crew might get some refreshment. Docking at the wharves, the company marched uptown for a round of drinks then marched back and departed. The anxiety they incited was like a match to tinder and remained lit for years.

Another of Sam Irvine's stories concerns cousins and brothers from the Civil War generation of his family who discovered a cache of gunpowder and abandoned firearms during a time of their general confiscation. Game had come back into the county, hunting being difficult without arms, and making do was hard; so with the powder found, those boys thought they might risk it. Often, however, four or five young men have little sense to spare, and these boys, while "recovering" the powder, started tossing chunks into a winter outdoor fire until, just fooling around, they blew the whole cache up. "This tale wasn't told till years after the war," concludes Irvine, underscoring their lasting fear of discovery.

I don't know how much powder went up in that farcical act, any more than I can calculate the extent to which feelings honed in that time still influence debates on gun control. But the phrasing of that debate, especially in rural areas—"the right to bear arms" as opposed to "gun control"—parallels my uncle's "War Between the States" and my father's "Civil War." And it draws on memories, for example, of ex-militia men who once lived in the bottoms, enjoying its cover.

If those men weren't bushwhackers, they acted about the same. They raided Southern sympathizers like the Irvines and ruled with a high hand. Though the war was over, these men still relished coming into Miami and shooting up the town, finding little more pleasant than striking fear in their neighbors. Enough guns had been confiscated during the war that the Southerners fought back on unequal terms.

Mart Rider was an ex-militiaman. Rider had signed up with a Confederate troop but after some months came home and reenlisted with the North. With the militia he could confiscate guns and horses from his neighbors, and since Southern sympathizers tended to come from the landowning classes, who had larger farms, hemp to harvest, and more slaves to manage, being able to turn the table on such folks must have felt pretty good.

After the war, Rider lived in the bottoms downriver from Miami with a wife from across the river to the north and four or five children, all of whom he was said to beat up on pretty good. He made a good bully and took a liking to patrolling a small bridge over a ravine just east of town. Like Billy-Goat-Gruff with a shotgun, he'd say, "No one passes today" or "You can't pass." A fellow came along who could not pass, because he could not pay, and Rider made him dismount, unsaddle his horse, saddle himself and get down on his hands and knees, and then granted him "permission to pass" by riding him across the bridge. So it is said.

Another story is that Capt. Joe Wilson came along, another sort of man. Wilson had ridden with Colonel Shelby, a man so dedicated to the Southern cause that rather than surrender he went on down into Mexico. Now after the war, Wilson too lived in the bottom, "making a citizen," as Sam Irvine put it, which seems to have been at least as hard as "building a book."

"Nobody passes today," Rider pronounced again at the bridge. Wilson just turned his team around and said, "I'm going home, but when I come back, I'm going to cross." Rider knew that meant he'd come back armed and decided, prudently, that "Captain Wilson could pass."

Like Wilson, plenty of Miamians knew where to turn for a gun, for emergencies, they said, or for a man like Rider. One day a band of ex-militia came out of the bottoms and into town where a Mr. Snelling had a store. A little black boy was sitting in the door. "Watch me pick off that nigger," one of the riders boasted. He fired, but his shot went high.

At that moment, Mrs. Snelling was stepping out of her store, and the ball struck above her head. Now Snelling was a hot-tempered man, and he went for his gun. How far he had to go, sources don't say. My guess is to the nearest drawer. Several of his neighbors seemed able to find weapons on a moment's notice, and the battle started.

The raiders backed out of town and headed toward the river, since the road was down that way. Guns opened up everywhere. One militiaman was killed, and another lay in the road acting mighty dead. A citizen ran up, saw the raider was alive, and started to repair that when a neighbor hollered, "Don't shoot a dying man." So he ran on with his friends, chasing the other villains back into the bottoms while the "dying man" jumped up and ran off like a bushwhacker.

Though the Miami men ran that bunch out of town, they had to charge somebody with murder since one man lay dead. So they picked a leading citizen and charged him. At first, no one was willing to defend the accused. Everybody was still that afraid of the militia. But finally a lawyer was found, and at the trial he walked in, took off his coat and laid it down. A big gun stuck out of his belt. "Gentlemen," he said, placing the gun on the table, "I brought this along just in case."

Rider may well have been among those raiders. In any case, he had a young fellow working for him who sympathized with Rider's young, beat-upon wife, and he offered to help her back across the river could they find the moment for it. So one day when a light snow covered the ground and Rider had gone up into town, this young fellow found a boat down by the river and rowed that young wife on across. Rider came home and saw that they were gone. Suspicioning what had happened, he tracked the fugitives to the river, then went home and got his gun, a muzzle-loader, but evidently he didn't have any powder. He went to a neighbor, a man by the name of Burns, or maybe it was Burnside, and asked for powder, said a hog was getting in his potatoes, though snow's not exactly the time for potato harvest. Maybe he meant that the hog was getting into where his potatoes were stored.

Anyway Mrs. Burns(ide), her husband not being home, tore off a sheet from the *Christian Advocate* and wrapped up a measure of powder for Rider. She made a little pouch of it as if she'd used her hanky. Rider loaded his gun, using the paper for wadding between the powder and shot, and lay for his man in the willows along the river.

Well they found the body later, and looking for evidence, found the wadding. When they spread it out, they discovered that it was torn off right across the name plate of Mr. and Mrs. Burns(ide). The pieces fit together, like impatience and a rope, and with that evidence, they hanged Rider. "One of the last legal hangings in Saline County," is the way Sam Irvine put it.

The photograph is dated 1888. Rider stands tall in a dark suit, his hands bound, the noose around his neck. There is no fear in his face that I can detect, nor repentance. He looks out on the crowd with a degree of defiance, perhaps contempt. The citizens have made it a festive occasion—parasols, Sunday dresses, picnic baskets. Little boys clamber on the gallows as if it were a jungle gym and mug for the camera.

One printed account adds that the piece from the paper was torn across the text of "Amazing Grace," right between "grace" and "wretch."

I've also heard on pretty good authority that wadding, when it falls from a muzzle-loader's barrel, tends to start fires in the grass, though perhaps, since it was winter, those sparks went out in the snow.

"Why, Mrs. Hamilton!"

After the concert, my grandmother approached him slowly, she and my grandfather in a long line with others. Boone, the pianist, was blind and black. Years before, my grandmother had been his teacher. Now Boone played on a Chickering grand piano made entirely of oak and had an annual income, before the First World War, of $17,000. A degree of myth had enveloped him. Program notes of the time summarize his story:

> Blind Boone was born in Missouri, Miami, Saline county, in the Federal camp, in 1863, 7th Militia, Company I, his mother being a contraband, cooking for the soldiers. He lost his sight with brain fever, when six months old. His first instrument was a tin whistle on which he could play any ordinary air after once hearing it. Next he was presented with a mouth organ, by which he charmed the whole neighborhood, children coming from far and near to hear him exhibit on his mouth organ. He soon became the favorite of all who knew him, and visited the best families in Warrensburg, where he makes his home at present. They formed such an attachment for Boone that he was sent to the St. Louis Blind School to learn a trade, and educate him. This was a failure, however. Once hearing a pupil in the institute practicing on the piano, he would leave his work and steal to the piano, as it was impossible to keep his fingers off the keyboard. He soon became able to finger out several pieces, and it was impossible to get his mind on anything else. He was dismissed from the school, and wandered around St. Louis, making his living by playing on a mouth organ, and such instruments as he could get his hands on. Conductor A. J. Kerry, seeing the pitiful condition of the boy, put him on the train and sent him to his mother. He soon organized a little company, and started on the road, tramping and beating his way from town to town. His company consisted of a player on each, a tambourine, triangle, and mouth organ, by which they gave concerts on the streets. He was not successful, however, and he endured many hardships. A colored gentleman, Mr. John Lange, of Columbia, Mo., taking a liking to him, put him in the Sunday-school to play for the children. He also made a contract with his mother to educate him in music and put him on the road, and he has made a grand success of it. . . .

Some sources say that Boone's mother was a former slave of descendants of Daniel Boone and that his father was a bugler with the Union army. "Contraband" was a term for women in his mother's position, freed or escaped slaves attached to the Union troops. "Brain fever" names encephalitis, and some accounts also say a Warrensburg doctor removed both his eyes surgically, since that was thought to be a cure.

Readers of Willa Cather will remember a similar case. In *My Antonia*, Blind d'Arnault, a black pianist, gives his concert in Black Hawk, her fictional version of Red Cloud, Nebraska. D'Arnault does not escape stereotype as a black musician, but several scholars say that he was drawn from life. James Woodress mentions "a real Blind Tom, whom [Cather] heard in Lincoln, and a Blind Boone, whom she probably had heard in Red Cloud."

D'Arnault, Blind Tom, and Blind Boone share a motif of discovery, of "stealing" to a piano that had been in recent use, not at a school in the first two cases but through an open window at a master's house where a daughter had been playing. Thus each is discovered to have an obsession for the instrument, which leads to their instruction and careers. Boone and Tom intersect again at a Blind Tom concert. Lange had taken Boone to the concert, and when Tom challenged the audience to play something he could not repeat from memory, Lange urged him to accept the challenge. Boone, then fifteen, did and won; the event launched his career.

Boone played in New York, Boston, Philadelphia, and Washington, at Harvard and Yale and made two trips to Europe. Rachmaninoff, Paderewski, and Ganz praised him. Old newspapers provide notices like this, from the *Miami News* of 1894:

> Mertens Hall was filled to its utmost capacity and hundreds of people were turned away on Saturday night, the occasion being the Blind Boone Concert. Those who attended enjoyed the music greatly and say it was by far the best entertainment given in Miami for many years.

Boone liked to say that he "put the cookies on the lower shelf, where everyone could reach them." Along with Gottschalk, Chopin, and Liszt, he offered his own versions of popular medleys and songs, southern black stereotypes such as "Camp Meeting," "Plantation Song," or "Suwanne River (with variations)." He was best known,

though, for his "Marshfield Tornado," through which he "reenact-
ed" an 1880 tornado that had all but destroyed Marshfield in south
Missouri.

From all accounts, Boone refused to record this composition.
Some say that once he attempted to record it on a piano roll but
"stripped the gears." Others declare his insistence that this piece, being "part of him," would die with him. Each performance was a variation, "beginning with the quiet of the Sabbath morning at church,
the approaching storm, and the roaring destruction as it sweeps over
the town. The noise dies down and peace again reigns. . . . down
to the quiet dripping of the waters." One account says "the young
musician used all the notes on the keyboard, and played with his
elbows, wrists and knuckles, as well as with his fingers." In Marshfield, people relived the tragedy when he played it; "some burst into
tears and dashed outdoors to examine the skies."

It became standard for Boone, as for Blind Tom, to invite a member of the audience to play something that he would have to repeat. In Marshall (rather than Marshfield), Eddie Lowenstein "volunteered and played a long classical number. Boone replayed it."
Whatever else influenced audiences, they had ample respect for
Boone's talent, which may have brought my grandmother into service as his teacher.

All the stories say that Boone learned the classical repertoire from
such a teacher. Several accounts locate her at Christian College, in
Columbia, Missouri, where my grandmother was professor of music, specializing in the piano. Some mention a professor at Iowa
State. Both my father and uncle, in correspondence to different in-
quirers, repeat their understanding that their mother taught Boone.
They offer no elaboration, and in one of the files, another note, by a
researcher, says their claim lacks substantiation. All I can add is that
I've never heard anything from my father or uncle to suggest that
Grandmother was given to boasting. From all I remember, I would
find it more remarkable for her to have invented the story than to
have lived it.

I used to visit her regularly. Her room was dark with heavy, hardwood furniture, the only natural light from a single window to the
east shaded by an elm. A small desk lamp was apt to be on, and
Grandmother often read by bending closely over a text, using a mag-

nifying glass. We played Chinese checkers, and she often let me win. Often too she seemed careless about one point of personal hygiene. She had a full set of false teeth, which she would soak overnight in a glass of baking soda and water. When they were not in place, her mouth sagged, and often I caught her in that condition. The teeth would stand isolated, on the corner of her desk, emitting a whitish glow in a glass from which I might have been offered milk. Perhaps it wasn't carelessness, perhaps she enjoyed my reaction, for she never voiced the least surprise or concern.

My grandparents attended a concert like the one at Mertons Hall. Afterward, they joined the receiving line and waited, "with farmer patience," as Grandfather told it. Emerging at last at the head of the line, Grandmother stood as quietly as corn for a moment, subtly interrupting the rhythm of welcome and congratulations. Then without a word she extended her hand. It was Boone who broke the silence, exclaiming, "Why Mrs. Hamilton, I'm so glad you came" the instant his hand touched hers.

The Hatchet

Nobuddy ever fergits where he buried a hatchet.

—Frank McKinney "Kin" Hubbard

One would like to hear Grandmother's story as a healing story, one of burying the hatchet. But the stories one collects are often at such cross-purposes that it is impossible to know. My father remembers her asserting that a Negro brain "ossified" at puberty and was therefore incapable of retaining any new knowledge. Where had the Chicago woman picked that up, and how had she retained it? My father took it to be an idiotic rationalization of Southerners, into which she had married, though he remembers no such words from his father. How does such bigotry speak for or against her work with Boone? Father also remembers there never being an underdog she didn't favor. She traveled to Germany between the two world wars on a Quaker-sponsored peace mission. She influenced her older son, my uncle, to serve with the Quakers. Later she wrote articles for a magazine that always ran a banner on its cover, "Prevent World War III," with a photograph of a mushroom cloud beneath it. Father assumed that the magazine was sponsored by "a communist front" organization. She subscribed for decades to the *Catholic Worker*, which kept on appearing long after she passed away. From all of which we may understand once more that contradiction is the shape of character.

I remember going to the University of Missouri for a football game while in high school and seeing the Confederate flag waved brazenly by students only a few years older than I and thinking that it seemed a little "tactless" but not yet finding it wrong. The Civil Rights movement was several years off; our consciousness had yet to be "raised," though it had been stirred. It took me longer to see a sorry relation between the long views of landscape that I cherish and the condescension, sometimes savagery, we practiced on them. How easy it is to overlook terrible things. I hope that tree did die, and if it did not, the thought, widely circulated, may count for something. I find it likely that Judge David Todd, so engagingly liberal with another man's slave, was adept at keeping his own man off a rope. That's beyond proof, but not above suspicion, and we may admire the result while questioning his motive. Or again, within half a year, James Shackleford moved away from the community he had lately led. To me that suggests a certain wavering in, and discomfort with, the stance he had exerted himself to express. So he got away from where his views were most vividly reflected back to him in the faces of neighbors. Or perhaps he just moved deeper into conflicted border country.

I do not know how to interpret Paul Stonner's account of the license given, or forced upon, some men, "if they wanted protection," to lynch another man who had killed a third man "who was awfully well liked"; but I see no reason not to call it coercion or to assume that "awfully well liked" and "didn't much like" would have squared with the black community's evaluation of those individuals.

Another murder in Miami, this time in the 1920s, found one citizen shooting another through the screen door at the barbershop. The murderer got off because of "an unwritten law" concerning fooling around with another man's wife. In this case, the murdered man was "awfully well liked," and the murderer was not. But he got a break that was not extended to the unnamed black or to Mart Rider, forty years earlier, who might have entertained similar suspicions.

Rider had all the makings of a youth we might seek now "to understand" and "to help." Most likely of the laboring rather than the landowning class, he got caught up young in a violent time. As sides sorted themselves out, he switched from Southern to Northern allegiance and joined the militia, from which vantage he could get back at a community that had long condescended to him. I'm guessing now, but that is my interpretation. He wound up a bully, a thug, around whom bad stories collected. When he finally went too far, any potential "understanding" was withdrawn like the trapdoor of the gallows, his sole support at that last moment.

The sense of celebration summoned for his hanging seems an attempt to bring mob behavior out into the sun. The women wore bonnets, and the children acted as if on a school-sanctioned field trip. Boys clowned around and provided comic relief. This was a "legal" hanging, after all—after all those others, and making up for process they had lacked. The community turned out in support and perhaps as a way of saying, about the lynchings, we were not so wrong then either—technically, perhaps, we were, but not in fact.

More than twenty years after the war, they were still seeking a scapegoat, which is precisely an attempt to coalesce and focus a range of griefs on one person and so to do away with them. George Mart (Martyr) Rider is not a bad name for a farm town scapegoat. That his grave still gave a thrill to boys a generation later, as my father's correspondent confessed, suggests that the citizens of Miami knew damn well where they had buried the hatchet.

Blind Boone, meanwhile, seems to have become a show horse and Rider's opposite. Boone was an individual so special that celebration of him would seem to ease the griefs long laid upon his people. A peculiar claim though concludes the program notes that presented him: "Boone has been on the road some ten years, has only lost about thirty days in all, and has been successful in pleasing his audience." The remark implies a working mule rather than a show horse. Apparently audiences took comfort in that. However much they admired Boone's ability, his audiences required a way to condescend and so to keep the upper hand.

Four generations of Hamiltons.

George Hamilton, 1981.

Anna Heuermann, age twenty.

Ted Hamilton about 1935.

Ted and Henry Hamilton in the Ozarks about 1935. Photo by their mother.

The homeplace in Callaway County after the tornado, 1921.

Clearing the farm, 1947.

Henry Hamilton with brush rake, 1947.

Magda Heuermann in the garden with her cat and dog.

Leonè and Ted Hamilton at Lake Michigan, 1940.

Henry and Jean Hamilton, 1980.

Copy of Spiro Mound copper by Jean
Hamilton and Eleanor Chapman.

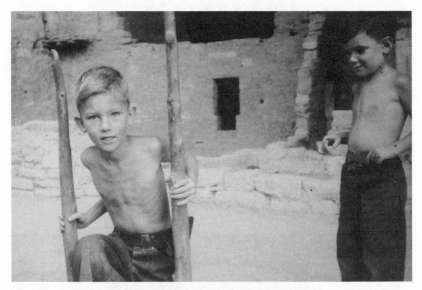

David and George Hamilton at Mesa Verde, 1947.

Ted Hamilton firing black powder for Miami schoolchildren, 1985.

Three generations of Hamiltons.

III
The Missouri Princess and Petit Missouri

One Boy

Corn harvest would drag on and on. In my father's youth
it had lasted all winter. Weariness would envelope him as he and
my uncle struggled to bring the corn in while the weather worsened
and the equipment broke down. But as combines outperformed the
old pickers, and newer, fast-maturing hybrids allowed harvest to
begin sooner, the work went faster; and though I can't remember
the year, I can still hear the satisfaction in my father's voice when
he announced that for the first time in all his life harvest would end
before Thanksgiving. It was like the arrival of a secret spring as he
anticipated an early return to his winter projects.

In the earlier harvests, that old two-row picker would poke into
standing corn clumsily and shake some ears off the stalks before they
could be caught on the rollers. From corn lying on the ground, small
thickets of "volunteer" corn sprouted the following year where they
had no room to develop but would suck up plenty of moisture trying. These clumps of corn and many single stalks besides required
hoeing out by hand. A cultivator could only catch a part of such a
clump, and any spray that killed the volunteer corn also killed the
good corn, and corn lost was money wasted.

Consequently we gleaned the fields. At this, my brother, cousin,
and I could help. Well into school by that season, we couldn't do
much else. The men could handle the picker and wagons without us.
But we could walk through fields, pick up fallen corn, shuck it, and
toss it into a wagon. We could tape "church keys" to the palms of our

gloves to use as shucking pegs. We could stomp through the fields, the better to shake the cold from our thin-leather-soled, ankle-high work shoes. We could keep our hands moving, having compromised on thin cotton gloves because they let our fingers find their way inside the shucks. And we could play.

The wagon sideboard made an obvious backboard, and I started shooting baskets. I backed off and tried a set shot rather than an underhand scoop, then banked in a hook. Once or twice I made a jumper, at some distance, while floating across a furrow. When I missed, I had to run after my shot and stoop a second time. George and Cousin Jim picked up on the game, and soon we were egging each other on and calling for passes.

Uncle Henry let this continue for a while, figuring perhaps that we'd tire out. But we had more in us than he was counting on, for soon he walked into our midst, turned on me most directly and observed, as if summing up all the evidence we had been laying out before him, "One boy's a boy, two boys are half a boy, and three boys are no boy at all."

With that in my ear, I've often stood a bit apart, trying to amount to at least one boy alone. One autumn not all that long ago, I rose early, pulled on boots, sweater, and jeans, and walked outside before dawn. To the east an old fencerow traced a line of palmettos, live oaks, and pine. Beyond the tree line, a marsh spread wide, with a bay close on the north edge; the marsh swung a mile or more to the east and then south another two or three miles. A second tree line marked the eastern, south-running boundary beyond which the Atlantic surged.

Along the edge of the trees, an uprooted cedar lay horizontal to the ground. Rootball and branches held the trunk as high as my chest. A hurricane had leveled it a month before, but the cedar lived on, green and vibrant. For weeks I spent hours each day on this point of vantage, with elevation enough to extend my sight and cover enough to remain unobtrusive. I studied how fritillaries, those phosphorous-backed butterflies, flame out and out on the salt marshes; how tree swallows cluster then vanish as if blown out like candles, then how, like trick candles, they reappear; how the heron, noticed, lifts and settles farther off; how his landing along a slight darkening means a rill within distant grasses; how the marsh hawk skates close to the marsh while the vulture, with the same dihedral set of wings, soars high overhead; how rails may be heard but not seen; how the vulture, so graceful soaring, becomes a bumbler trying to land; how the chubby, perching kestrel lengthens to a Lancelot in flight. Below me, fiddler crabs frisked in and out of holes as the tide ebbed and crept back.

I remember wondering why I was never bored, standing, staring marshward, observing one portion of the world, and I supposed that working alone, often standing at the wheel of a tractor, had prepared me for this long view not wholly dissimilar. The tractor had a seat, but Uncle Henry stood. On the John Deere, he usually stood in front of a padded seat on which he could have rested. On the Minneapolis-Moline that we bought from the Chevaliers after Bud was killed, he stood to the side of a higher seat of metal, saddle-shaped with upturned iron edges. Even when we hoed corn or beans by hand and the rest of us hastened, after two or three hot rounds, to sprawl under a cottonwood and drink from a water jug stashed there, Unc stood at the edge of the shade, tilted back his fedora, filled his pipe from the pouch in his work shirt pocket, lit it, leaned on his hoe, and stared back out over the field.

On the tractor, Unc stood straight, his worn fedora shading his eyes, one hand on the wheel, the other resting on his hip, his feet spaced for balance, and he rocked with the tractor as it picked its way across old furrows that had settled back into each other. The big wheels on either side held him high. I imagined him standing higher on them, straddling the two, stepping as they turned, and by a kind of moonwalk striding hugely across earth that ran back beneath the tractor to meet the plow.

Spring would barely peek over the horizon and he was in the field. In the late fall and early winter, he remained there, plowing under stalks of recently harvested corn. My father once said that "the trouble with Henry was that his idea of play was another full day's work," and there was truth to that, though Unc's attitude toward work was often playful. Moreover, he seemed best attuned to work well aligned with a boy's idea of play. If I'd tossed him an ear of corn that early winter afternoon, he might have shown me his set shot. And when have you known a boy who did not want to take the wheel of a tractor?

I'd go to the field expecting a day of slow but steady escape to stretch before me, a day of little distraction except the occasional stop to pull refuse from between the plow bottoms or the cultivator shoes when those became clogged or to change a shoe if one broke on an old stump. I'd climb high to sketch out the stories I favored, to lay out my trapline and construct my cabin in the North Woods, or, like an aspiring Missouri warrior of the Buffalo clan, to prepare for a long hunting season and my vision quest. Often I wondered whether a young mixed blood, Petit Missouri, had got past my age on the plains.

Possum

An old family photograph shows me emerging from a kiva at Mesa Verde National Park, climbing a ladder from an ancient ceremonial chamber representing Mother Earth and a preceding world into the succeeding one of our days. I'm shirtless, skinny, self-conscious, and eight. George stands to the side, ready for his turn. It was during our first summer in Missouri. A flood had brought all clearing and farming attempts to a halt, and my father and mother took refuge in a family vacation to the Southwest, a region of which they had read but never visited. We saw Anasazi ruins in Mesa Verde, Hopi and Navajo lands, witnessed a rain dance, and came home with a wooden kachina that still overlooks my work. Our parents packed us into their Nash, stuffed camping gear into the trunk, and carried woven water bags along, both for a possible overheated motor and for crossing deserts, should our passage take longer than planned. We were gone a month, by far the longest family vacation I remember, of which there were only a handful anyway since farming soon overwhelmed our summers; and I was told, when I raised the question later, that the whole trip cost two hundred bucks. But the dollar was a different dollar in those days, as legions of our forebears have testified, and perhaps in the long run it was cheaper to get us all out of the house.

My strongest single memory is of kivas, of being permitted to climb down ladders into several and to feel their intimations of mystery and deep memory foreign to me but long conveyed to appropriate others by rites underground. In that outdoor summer, when we watched stars at night and emerged from a tent each morning, I sat shirtless, with my bare back pressed to cool earthen walls and tried to imagine chants, dancing, and the instruction of boys like myself by men like my father and uncle in the essential lore that was the ancient cultural information of the Anasazi. I put my ear to the *sipápuni*, the small hole in the floor representing the place of emergence from a still-earlier world and through which spirits could pass. Not that they would traffic with me, an outsider, ignorant and uninitiated, but I dared to imagine their indulgence.

Archaeology was becoming the family hobby. Books of the West and of Indians were among my treasures as were several old copies of *National Geographic* that depicted pre-Columbian life in this hemisphere. Artifacts from Saline County would soon line my shelves and drawers. As the years passed, I pored over my arrowheads, knife blades, and smooth slabs of hematite. I loved to hold them in my hand, to study the colors of the flint, the delicacy of the chipping, especially those with beveled edges slanting in opposite directions. The hematite, a smooth red stone, seemed capable of magic.

An older man who had lived among the Sioux came to town and joined our arrowhead hunts. He gave me a beautiful Sioux tomahawk that could only have been ceremonial since its superbly rounded, double-pointed head was bound by buckskin and set on a slender shaft. Though it seemed fully capable of murder with one blow, I doubted its capacity for staying bound to the handle for a second. It was a gorgeous piece of craft, however, its granite head symmetrical, grooved for the buckskin binding, polished and mottled in pinks.

On the farm, the uncleared strip of timber called to me whenever I had free time, and I maintained several paths through it by the frequency of my explorations. In town our street was the highway running eastward out of town, with no other streets or development behind it. I had about a hundred acres of abandoned pasture and forest largely to myself, with paths, glens, and hideouts, with animals and birds, and with all manner of places for gaining a woodland perspective on home. For however domesticated the farm was becoming, in fact, and no matter how truly I lived in a snug frame house with a room of my own in town, the out-of-doors remained an overwhelming part of my life, and Indians, all Indians, no matter how romanticized my imagery, no matter how maimed by history that I ignored, were the unchallenged authorities of that world. With respect to the culture of the out-of-doors, the Indians were my classics.

A grove of Osage Orange, a favorite Indian bowwood, crested a knoll north of our house and was my first hideout. Friends of my father who crafted fine bows as a hobby made a longbow of Osage Orange for me that I carried into those pastures and woods. I practiced stealth, avoided stepping on twigs, and tried to believe that I could slip into the trees and disappear. I conflated Robin Hood with the Indians, for the outlaw romance was another of my favorite books, by which I mean one I read over and over, preferring that to starting something new. The bow of the woodland Indians was also a longbow, as my father, who was then beginning a study of "Native American Bows" knew well and would elaborate further. I made arrows for hunting rabbits and fletched them myself, but don't remember ever hitting a rabbit.

Then there was Dallas Gordon, an early hand who lived in a shack on the farm, hunted the strip of timber himself, and while we hoed corn, he taking eight rows to my two, which later grew to four, recounted his boyhood in Indian Territory. I have no idea how much he made up and how much was true, whether Dallas was part Indian himself or for other reasons had lived among them in Oklahoma. He disappeared from our lives soon after. At the time, though, he seemed an informant beyond my father and uncle whose more studied knowledge remained secondhand.

Dallas told of sweat lodges, of how they were constructed of bent saplings, buffalo hides, and wooden pegs, of the heated stones heaped in the middle and of crouching inside, naked or nearly so, of sweating until unable to bear the heat any longer—for longer, he insisted, than it took us to make our mile round of the field, which would have required more endurance than I could imagine—then of throwing back the flap of the lodge and running to the cold pool only a short sprint away and hurling himself into it. He described the darkness inside the lodge, the constant drumming and singing of elders. After a time, you couldn't make out companions on the far side of the lodge but hoped to last at least as long as they. The smell of cedar bark, thrown on the stones, distracted you from your discomfort, and gradually you turned inward and heard, or thought you heard the rhythms of your own breath and heart. You intuited without being told that the sweat lodge, like a kiva, was a womb, and that when you ran from it, into the river, you were reborn into a more crystalline world. The extremity of the sensations and the abandon of it appealed to me. So did the nakedness and the hints of reaching beyond myself. I remembered the cool earth on my back in those kivas. Hints of the erotic awakened in me like the eyes of small animals at night.

I was preadolescent or just turning. Girls had entered my mind, but most things shared I still shared with a few boys who were close friends. Sometimes one or two of them would come to the farm, and we would camp in our strip of timber. Mother helped us make breechclouts. These are summer memories, which always meant much shedding of clothes. We constructed wigwams, built fires for rudimentary cooking, gathered mulberries for painting our skin, and made a few raids on the barn. We were coming into our own bodies then, and the near-nakedness of Indian life as represented in our books led our way. *National Geographic* has been known to serve as a first pornography.

One of my friends, Johnny, and I imagined slashing our chests, looping thongs through the wounds, and lashing ourselves to a tall, sun dance pole. We had read of a torture by which a victim would be tied to strong, young trees that had been pulled down for the purpose and that would then rip him apart when they were released. I felt the thought of that in my groin. Stripped to our breechclouts, we experimented, decorating our exposed chests with mulberries. Then one of us bent the other over backward and lashed him to youthful soft maple and box elder. As the torturer released the trees, they would pull on the victim and emphasize his temporary exposure. But we knew our trees well and knew those were incapable of inflicting real harm. Ideas of Indians, though, animated us both; many of them have lingered with me through whatever further understandings I have acquired.

A few years ago, Hyemeyohsts Storm came to town. Storm is Cheyenne, perhaps a shaman, and the author of several books that convey Cheyenne views of the world. *Seven Arrows* made him a cult figure and probably remains his best-known work. He was visiting a colleague and friend who called and said, "Hurry over." I had

no idea who awaited me, but no sooner had I entered the room—
I barely stood within the frame of the door—than Storm turned and
said, "He's Possum."

Years of make-believe swirled back. Yes, I'd rather have been
Wing-on-the-Wind, or Storming Raven, or Coyote Dreaming, but
self-naming was not his offer. I was just being introduced, in fact,
under the name by which I normally pass, when Storm continued
blithely, "you don't find one often." His confidence kept me silent.
The conversation drifted to other things, but when I got home I
pulled out a few books and read a version of myself to which I'd
paid small heed. A possum, though occasionally encountered, had
never seemed fully worthy of my attention. To Indians though, at
least when remembering their better ways, nothing of nature is ever
beneath one's dignity.

"Small brained," I read, "instinctual, many teeth, chews things
over, nimble at night, a survivor, likes sweet streams better than
roads, seldom follows the same track exactly."

"One boy," I thought, and supposed I could live with that.

Soon after that evening I got out the fine old Osage Orange bow
that those friends of my father's had made for him about the time
they made mine. My father's had twice the strength of pull of mine
and would bring down a buck were I a hunter. My old arrows were
ratty, so I deigned to buy a half dozen target arrows at an indoor
range north of town. Then I rented a lane, rather like a bowling alley,
and strung my bow.

Archers on either side of me paused and stared as if I had stepped
from the thirteenth century. Their bows were more laced than strung,
with pulleys and tabs that monitored the length of each draw. Almost like rifles, some had sights! True enough, they grouped their
shots better than I. But all but one of my arrows also flew to the inner
two circles of the target, and two stuck out smartly from the red.

People of the Canoe

It is an irony twice over that the Missouri River, one of the longest
in the world, as capable as any of carrying us deep into the past,
combining with the Mississippi to drain an area rivaled only by that
which feeds into the Amazon, drawing into it dozens of tributaries,

creeks, and smaller streams and providing over half the volume of the Mississippi as it passes St. Louis, should be named for an obscure tribe, now extinct. "Oumessourit" was what Marquette and Joliet transcribed from what they heard the neighboring Illinois give as a name. It has been translated as "the people of the canoe" or as "people with dugout canoes."

Since the Missouri had split off from the Winnebago and migrated from the birch forests of Wisconsin to the cottonwood- and willow-fringed river they would name, they had proved themselves adaptable and competent. They learned to shape a walnut log into a craft stable enough for the big river and passed their learning on to trappers whose iron axes and adzes enabled them to refine the vessel. Walnut endures indefinitely in damp soil or water. It's no surprise that a walnut log was found at the bottom of the Grand Pass spring. A trim twenty-four-foot walnut craft found submerged in mud in northern Missouri closely matches one in George Caleb Bingham's painting "Fur Traders Descending the Missouri."

But the Missouri lost their own name along the way, which was "Neutache" or "Nu-dar-chee," for "those who arrive at the mouth," of the Missouri, perhaps—before they worked their way upstream— or of the Grand, a significant tributary beyond which they settled into a town that endured for over three centuries. Four miles south of Miami, the Missouri town site overlooks the Tetesaw and Van Meter bottoms.

Estimates of the Missouri population at the time of European contact vary from 1,000 to 5,000. Only two towns in present-day Saline County have populations of more than 2,000, and neither has its own language. As the first tribe to be reckoned with on the Missouri River, it could prevent trappers and explorers from continuing upstream. Bands that included the Missouri, Osage, Oto, Ioway, Pawnee, and Illinois ranged from Detroit to north-central Nebraska, and so perhaps farther. Finds at the Missouri site indicate a trade territory that reached from the Mississippi delta to Canada.

According to an old and simplified report, they divided the year between leadership by the Buffalo and the Bear clans. Other accounts number as many as seven clans, but the year began under the Buffalo with planting season followed by hunting on the plains. Fall and winter were the Bear's seasons of harvest and hibernation. From their situation on the bluffs, they planted corn, squash, and beans and set out on summer hunts while the lower growing plants kept weeds from choking out the corn. They ranged far on the prairie, their hunting territory intersecting with that of other tribes. Always they were on guard against skirmish and loss, and I often wondered, when working our fields alone, how I could both work and guard against ambush if Joe Chevalier, who was surely okay but not a blood brother, were to rally his brothers against me.

For it was as if "the people of the canoe" meant, more precisely, "those who float in their caskets." The story that unfolds is of disaster. When Marquette and Joliet learned of the "Oumessourit," they promptly claimed for France all the land the river drains. Before long, traders ventured up "the river of the Missouris," and voyagers had probably made their way upstream even earlier. In 1680, two were captured and taken to the Missouri town, which, as first on the river, buffered other tribes from European encroachment. Small pox and other imported diseases were the first killers, more devastating than centuries of intertribal skirmish and warfare, and more indiscriminate too, attacking Missouri leaders and the strongest among them as indifferently as the rest.

By the 1720s, estimates suggest their population was down to 1,000. In 1804, Lewis and Clark estimated no more than 300 and pointed to "the fury of the Saukees" as responsible for having recently killed an equal number. Given their reduced numbers and many leaders lost, they were much more vulnerable than before. Survivors melded with the Otoes and Osage. As white settlers pushed into the region—Daniel Boone and his sons arrived even before Lewis and Clark—their records mention the Kickapoo, Winnebago, Sauk and Meskwakie, Osage, Miami, and Shawnee, but there is no Missouri tribe left for them to confront. The last full-blooded Missouri is said to have died in 1908. Before those losses, however, our farm had been a fragment of their realm.

Like our upland neighbors, they would have looked down on us. I could see the edge of their bluff town site to the southeast of our Front Field, perhaps three miles away as the crow flies. To the southwest, just far enough off to grow hazy, the Little Osage had had a town of their own through most of the eighteenth century. During winters they probably all migrated down into the wooded bottoms.

The timber we cleared served as a windbreak for them, miles deep in places, and provided ample fuel.

The strip of timber left uncleared on our farm ran about a half a quarter wide and a half mile deep. My father and uncle left it to remind us, and to remind themselves, perhaps, of what they had done. They had also run low on expenses for clearing and on the will to wrestle into production every last piece of land. It always encouraged Indian consciousness and Indian play.

Trails I found in it mystified me. Were they old hunting paths of the Indians who had ranged through this territory? Were they trails early settlers had made, or bushwhackers of the mid-nineteenth century? Perhaps they were trails the deer took in the evenings on their passage to water. I saw squirrels often, raccoons now and then, and a bobcat once, disappearing, as elusive as a dream. Whatever their origins, I took them over as mine and ran them, dodging low branches, leaping over logs long felled by wind or flood or age and now decaying patiently in my path. Much of that timber had gone into our soil, layering itself into the earth for centuries. The word "timber" derives from the same ancient root as "domain."

But who were the Missouri warriors? And who was Petit Missouri? In a cigar store backroom in which men gathered to play cards and boys entered cautiously to hang out, a spectacular depiction of Custer's Last Stand hung above a cooler. Yellow Hair held the high ground, a revolver in each hand; dead horses formed a barricade behind which his soldiers leveled their rifles at the Sioux and Cheyenne. I saw movies featuring Sitting Bull and Crazy Horse, Geronimo and Cochise, but no Missouri warrior emerged in our folklore whether accurately or inaccurately.

There is a single exception to all this, or half an exception. Among the historical murals in the state capitol in Jefferson City, one tucked high in a dimly lit tympanum over double office doors portrays a young woman in an off-white gown with white pinafore and bodice. Her scoop-neck gown with three-quarter sleeves fits tightly to the waist, and she wears a broad, white choker. Her matching low-crowned hat is wide brimmed and bedecked with white flowers. Small, sharply pointed shoes peek from beneath her flaring skirt. Her fine features suggest a belle, although a dusky one. Her slender forearms contrast becomingly with her hat and gown of white on ivory. From beneath the wide brim of her hat, two long dark braids fall forward to her waist, one over each shoulder.

Six Indian braves, each a head taller than she, and darker, range in a close arc behind her. Each is bare-chested, although the one at her right hand and farthest forward has an off-white robe over his shoulders that echoes her gown. Each warrior wears a scalp lock, painted bright white against a base of red. Another man, barely taller than the woman, stands at her left inside the arc of the braves. He wears a tricornered hat, a green coat, and a white shirt with a ruffled front. She is the bride and he the groom. She is being greeted by another tall warrior who has the same white-on-red scalp lock and the ornamented robe of a chief.

The painting, which dates from 1923–1924, was a deliberate attempt to get it right, and shows that authenticity comes by stages, and work that satisfies may well be found wanting soon enough. Ernest L. Blumenschein and several fellow members of the Taos Society of Artists were commissioned for this and related works because they were "familiar with frontier and Indian life." An official report dated 1928 calls the composition "most effective" and calls special attention to its color scheme of "subdued harmonies." Though still too Hollywood for my taste, it portrays no plains warbonnets.

Three braves kneel at the chief's left hand. A patch of river appears in the background and a glimpse of the bare bluff opposite. Figures in darker clothing buttress the central group as they lug supplies up from a boat. The whole party has mounted the bluff on the south bank of the river. Pennons wave overhead. The climb has not left the bride breathless, nor has climbing from the river bank to higher ground made either her shoes or gown the least bit muddy.

The mural, "Return of the French Officer and His Bride to Fort Orleans," imagines an event of 1725. The young woman, "the Missouri Princess," is the daughter of the Missouri head chief. Along with lesser chiefs of the Missouri and several related tribes, she had voyaged to France, where they had been presented at court and before the Company of the Indies, both in native ceremonial costume and in French finery. They were given stylish clothing and other presents and were briefly the wonder of Paris. Later, she was baptized at Notre Dame and married to a Sergeant Dubois, who had accompanied her from Missouri territory.

About their return, reports differ. Some emphasize the joy the French took in returning a Monsieur and Madame Dubois to a new French fort on the floodplain near the town of the Missouris. Madame Dubois was said to have brought the first French gown into the state and great joy to her tribe, all of which the mural suggests. Then she took up residence in the fort. But there was little honeymoon, since as these same reports claim, the boat that brought them had hardly turned around than the Missouris massacred Dubois and the entire garrison of Fort Orleans, and their princess renounced Christianity and returned to her people and to her former way of life. In this version of the story, the princess is the heroine of a captivity narrative in reverse and prefigures Huckleberry Finn's escaping the widow's home and returning to his riverside haunts.

In other reports, Sergeant Dubois, who in the years before had traveled overland into what is now central Kansas and had voyaged up and down the Mississippi and explored the Missouri as far as the mouth of the Platte, was ambushed and killed. But this happened along the Mississippi before his party even reached the Missouri. His princess survived the attack but never made it back to her village on the bluffs overlooking the Tetesaw Bottom between Laynesville and Miami. She remained in Illinois country, where she remarried a captain of the militia and twenty-seven years later showed a French traveler a diamond-studded watch given her by King Louis XV.

Bourgmont

A young man named Bourgmont left Normandy in 1698, accused of poaching on monastic lands. In 1706, he turned up in command of Fort Detroit, then under siege by Indians. Already hardened to the wilderness, he had been up and down the Ohio and overland from Quebec to Detroit. Soon a battle occurred and two Frenchmen, one a priest, lost their lives. Held responsible by some, Bourgmont deserted. A subordinate who also deserted was captured and sentenced "to have his head broken till death follows, by eight soldiers . . . they being unable to inflict either a more severe or more moderate punishment because of the remoteness of the place." And because, I would guess, of need to hoard powder and shot.

Of *une famille vivant noblement,* that is, ambiguously, of a family who lived in a noble manner, Bourgmont either avoided capture or was allowed to slip away. Apparently Cadillac, in whose name he commanded at Detroit, and who would become his commander again, recognized him as useful.

Movement and more movement, across oceans, through forests, up and down rivers, across more than half a continent, and all of it centered—for our purposes—on the Missouri town south of Miami. Bourgmont slipped away from Detroit in the company of Missouri Indians who had come to assist their allies, the French. Frontier rumor claimed that he fell in love with the chief's daughter, the Missouri Princess, and she led him into her country. Then in 1714, having lived in Missouri territory for a time and reunited with Cadillac in New Orleans, he won a new assignment. Bourgmont led a party from the mouth of the Missouri to the mouth of the Platte in what is now eastern Nebraska. He kept a log of the trip from which a Parisian cartographer first mapped reliably that much of the river.

Routte qu'il faut tenir pour monter la rivière Missoury ("The Route to be taken to ascend . . .") prefigures by almost a century the *Journals of Lewis and Clark.* But as a ten-page document, devoid of ornament, it makes only a rough start.

> Saturday, [March] 31. Proceeded to the east; an island about a quarter of a league long; the river channel comes from south-southwest. On for about a quarter of a league.—South a quarter of a league.

A league is not quite three miles, an arpent about an acre. A month later, Bourgmont composes one lightly colored exception, less than a sentence:

> Monday, [April] 30. West one league.—West half a league.—Southwest one league; toward the start of this stretch one finds a channel coming from the south-southwest, which leads to the village of the Missouris. The island might be about a quarter of a league long; afterward, another small island and some little trees, above which one can see the village, situated on a lovely prairie. . . .

That channel and those "little trees" suddenly have a homey feel. They are close to the writer, and "the village situated on a lovely prairie" lay at the east end of the Tetesaw Plain, repeatedly praised by travelers and newcomers.

Lewis and Clark would remark on this setting eighty years later, when the town had been long abandoned and they could make out little of its site from below on the river. Not long after, Capt. Thomas Becknell tells of crossing the "Petit Osage Plain" overland on a journey to Santa Fe (1821):

> The traveler approaches the plain over a very high point of adjoining prairie; suddenly the eye catches a distant view of the Missouri on the right, and a growth of lofty timber adjoining it about two miles wide. In front is a perfectly level, rich and beautiful plain of great extent, and diversified by small groves of distant timber, over which is a picturesque view of nearly twenty miles . . . bounded by the fanciful undulations of high prairie. Description cannot do justice to such a varied prospect, or the feelings which are excited in beholding it.

Report, however, had preceded Becknell and Lewis and Clark, and Etienne de Véniard, sieur de Bourgmont, explorer, adventurer, and awardee of the Cross of Saint Louis for his long service in Missouri country, was the first to make it.

Bourgmont returned to France in 1720 and again in 1725. On his first return, he married a distant cousin. In time they had three children. On his second, he was raised to the nobility. So he saw his Indian wife married to his sergeant and stayed home, hoping to enjoy the fruits of an energetic quarter century. Bourgmont died less than ten years later in the Normandy village of his birth. With him died the title for which he had labored, for his French children all died young; his infant son and heir preceded him by two years.

His only son, that is, except for "Petit Missouri," born of Bourgmont's Indian wife. In 1720, Bourgmont had taken Petit, who was about six at the time back with him to France. No wonder his *Routte* has that homey touch. It was written in the year of the boy's conception, or of his birth. But Petit Missouri does not make the party for the second trip to France, and so at ten or twelve, he is separated from his father, as well as from his mother, and all but disappears from the record.

On his first return to France, when he took Petit Missouri and married his cousin, Bourgmont received the Cross of Saint Louis and was named commandant of the Missouri River. He returned to America with four assignments: (1) to get as close as possible to the Spanish for trade; (2) to open the way through Plains Apache territory, so far blocked to them; (3) to establish a defense post on the Missouri River and "to choose its location with great care"; and (4) after establishing the fort and organizing trade on the plains, to invite a number of chiefs to France "to give them an idea of French power."

Bourgmont's careful choice for the location of Fort Orleans expresses his affinity for the Missouri. Although its exact location remains unknown, a probable site is about five miles west of the Missouri town with our farm on the line between them. Soon the Missouri, in their last independent stage, moved to a riverbank site to be closer to it. The fort spread across an eminence near a creek called the Wakenda, meaning the Great Spirit, the highest power, the alpha and the omega, in the languages of the Missouris, the Otoes, and the Osage. The French desired trade with the Spanish and hoped the Missouri River would connect them, but they feared Spanish incursion. The colonial powers vied for Indian support.

Meanwhile, Bourgmont moved easily among the Missouris and their allies, wielding ample prestige and power. Once he reports refusing to receive representatives of the Otoes and Ioways until they promised to break an alliance with other tribes that he thought harmful to the French. Another time a Missouri warrior slew a Frenchman, and Bourgmont "had him killed in the middle of his own village by having his brother shoot arrows into him. To show there were no hard feelings, the brother sent me a nice present on my arrival."

Love is rarely a stranger to power. In an old Winnebago tale well within the linguistic heritage of the Missouri, a chief's daughter marries an orphan. Because of her position, the daughter suffers from a certain isolation, and she has her eye on the orphan who, though overlooked at first, proves a remarkable hunter. "Why didn't you court me earlier," she chides when he finally gets around to it. Such a tale may have prepared for the encounter of a Missouri chief's daughter with Bourgmont, perhaps an orphan in her eyes, parentless so far as she could see, but an able stranger from distant, unimaginable France.

Sunsets on the Tetesaw Plains rank high enough on the rosters of sunsets and romance. Summer lingers so long that one can imagine "Indian summer" having been coined at the spot, not as a rudeness but as tribute to the faithful recurrence of a gentle season. Springs that quench one thirst generally quicken another, and as Becknell anticipated in his description, the "feelings which are excited in beholding" that landscape may lead to dalliance most pleasant. According to legend, the Missouri and Oto tribes divided long before contact with Europeans when the Missouri chief's daughter fell incautiously into the arms of the Oto chief's son. So the Otoes moved upriver and farther west and called themselves "Wah-toh-ta-na," variously translated as "those who copulate," "lovers of sexual pleasure," or, more bluntly, "aphrodisiac."

Oto and Missouri lovers may well have clung to each other on the bluffs overlooking the river. How odd to think it unlikely. Then Bourgmont and his Missouri Princess, generations later, took their turn under stars more numerous than the campfires of all their ancestors' ancestors. Perhaps the Frenchman pointed out the North Star and Ursa Major. But Indian traditions and archaeology both suggest that the Missouris had migrated south from origins near the Great Lakes, so perhaps the Missouri woman saw the story of her people in the stars and told Bourgmont that the fixed star was "the place from which my people came," and the dipper "the path we followed to get here."

Bourgmont brought news of a world about to engulf her people. His princess was well placed, young and vital. The westernmost toehold of the French at that moment took shape from their attraction.

Massacre

In 1720 a Spanish expedition from Santa Fe is said to have probed Saline County, intending inroads on profits accruing to the French. Bourgmont and Petit Missouri were out of the way that year, having sailed to France. The Spanish knew the Missouri to be longtime allies of the French and so wished to destroy them. Accordingly, they proposed an alliance with the Osage, "the ancient enemies of the Missouri." Coming upon the prominent Missouri village on the edge of the Tetesaw Plain, they mistakenly thought they had encountered the Osage and proposed an alliance against the Missouri. The Missouri chief agreed and summoned his council. The Spanish armed the Missouris and went to sleep pleased with their diplomatic triumph, but the Missouris mounted a surprise dawn attack and annihilated the Spanish, except for a priest whose unique costume displayed either his great medicine or his comic value. Indications of slaughter near the Missouri site, including shot and a Spanish saber, fortified one more tale told ever since.

As, with variation, it has been told elsewhere. Cather's Antonia, while picnicking on the bank of a stream, urges her friend Jim to tell the other immigrant "hired girls" of the Spanish expedition as he has often told her. Jim explains that they had been taught in school that Coronado had failed to reach Nebraska; Kansas was the limit of his march. But Jim knew of a farmer to the north who, "when he was breaking sod, had turned up a metal stirrup of fine workmanship, and a sword with a Spanish inscription on the blade." "I saw with my own eyes," Antonia adds in her imperfect English.

Historians now assume that an expedition into Nebraska did take place, led not by Coronado but by Villasur, and that it was destroyed on the Loup River, north of Red Cloud, which is the setting for Cather's novel. The attackers were Otoes and Pawnees, allied with and assisted by the French. Missouri warriors probably participated, since as a Missouri chief reminded Bourgmont later:

> We love the French nation and naturally hate the Spanish and English. We proved as much, three winters ago. The Otos, Skiri Pawnees, and we completely defeated a large party of Spanish who had come to establish a base among us.

The discoveries Cather knew of may have been relics of that battle. On the other hand, the Missouri may have been more involved than we know. Jean-Bernard Bossu, the mid-eighteenth-century traveler who tells the first version, as if it happened in Missouri country, also reports a strange and comic procession in which Missouri chiefs and warriors arrayed in chasubles, maniples, and stoles, and making ornaments of a paten and of "a chalice [hung] around a horse's neck as though it were a bell," presented a fine Spanish horse to the French commander in St. Louis.

For reasons such as these, the French established Fort Orleans. No sooner was it well begun than Bourgmont hastened to extend his influence to the Plains Apache. Through the summer and fall of 1724, one hundred Missouri warriors commanded by eight war chiefs and their head chief, and sixty-four Osage commanded by four head chiefs, along with women, children, and dogs, set out to accompany the twenty-five Frenchmen who needed to penetrate another two hundred miles onto the Great Plains. Rather than ancient enemies, the Missouri and Osage were kindred tribes who spoke nearly the same language and had long lived near each other. One wonders about the alleged Spanish strategy, even if related tribes often did, or could be coerced into, attacking each other.

It seemed a whole town on the march. Dogs dragged laden travoises. Women and children shouldered staggering packs, with girls of ten and twelve carrying loads that amazed the French, "but it is true that they can carry such a load only two or three leagues," the French report adds. The entourage set out every day before dawn and marched for five or six hours, then rested almost as long before advancing another several hours before dark. Sixteen miles was a typical day's journey through the heat of a plains summer. Hunters harvested bison and deer along the way—"twenty deer today and several turkeys"—while the women built temporary shelters and cooked. They fetched water for and mended the footwear of the French. Astonishing amounts of trade goods had to be carried and parceled out to the Missouri and Osage who set out with Bourgmont, to the Kansa, Oto, and Skiri Pawnee who joined the progress, but especially to sway the Plains Apaches, allies of the Spanish.

> When that moment came on October 19, 1724, Bourgmont laid out one pile of fusils, one of sabers, one of pickaxes, one of axes, one of gunpowder, one of balls, one of red Limbourg cloth, another of blue Limbourg cloth, one of mirrors, one of Flemish knives, two other piles of another kind of knives, one of shirts, one of scissors, one of combs, one of gun flints, one of wadding extractors, six portions of vermilion, one lot of awls, one of needles, a pile of kettles, one of large hawk bells, one of beads of mixed sizes, one of small beads, one of brass wire, another of heavier brass wire for making necklaces, another of rings, and another of vermilion cases.

Then he "harangued" the Apache chiefs and offered the goods. Their head chief accepted for his people and accepted the responsibility that went with the presents.

> [The Spanish] come to visit us every spring. They bring us horses and a few knives and some awls and axes, but they are not like you, who give us a quantity of merchandise such as we have never seen before. We are unable to reciprocate for so many presents.

But it wasn't as easy as that. Alarmed en route at the outbreak of a fever, and no doubt remembering earlier ravages of disease that often accompanied Europeans, the Osage returned home. Weakened and often purged himself, Bourgmont was transported all the way back to Fort Orleans, having accomplished only half of the proposed trip. He sent ahead a young man named Galliard who must have been competent for he continued with three Kansa warriors and two Apache slaves who were being returned to their people in an act of goodwill. Since the Kansas and Apaches were traditional enemies, maintaining peace deep on the plains required tact and vigilance. As the party approached the Apache site, the Kansa warriors fled and had to be coaxed back. But Galliard made contact, and Bourgmont returned to his quest. When he met the Apaches in October, Bourgmont established a brief "Pax Gallica" on the plains. Soon he returned to Fort Orleans and to France, taking representative chiefs and the Missouri woman with him.

Within three years, the French would abandon Fort Orleans. Early stories abound of its being overwhelmed, like the Spanish party on the plains. But France's investment in Louisiana had sagged, and its influence was ebbing. Very likely the fort was never completed. French government in the region, located in New Orleans, had much more pressing problems and what support they had managed had always been half-hearted. Bourgmont resided there less than a year. Lewis and Clark found no trace. Like the Missouris and other people before and since, the French lingered in then vanished from the valley.

First though the Kansa tried to solidify an alliance with Bourgmont by uniting him with a chief's daughter. Bourgmont declined, saying "Frenchmen were not permitted two wives." "But you are a chief," replied the chief of the Kansas, eyeing the Missouri woman with him. French chiefs, Bourgmont countered, had to set good examples. "Since you will not marry our daughter, we give her to your son, and he will be our head chief, and thus you will be our true father." Bourgmont could put that offer off. He explained that the ten-year-old Petit Missouri was too young to marry; perhaps later. A year later, father and son parted for good.

Our Discipline

Carl Chapman's handbook *Indians and the Archaeology of Missouri* includes a photograph of Chapman, my father, and me. We kneel in a well-excavated space at least two hundred feet square. Chapman, in pith helmet and jacket, holds a small flint point, which he shows to us. I would have called it an arrowhead, though, given its size and weight, it was more likely crafted for a spear. Kneeling beside my father, I mimic his pose: one hand holds a knee to the ground while my elbow rests on the other knee raised. Our outside arms make parentheses; our raised inner knees and elbows almost touch.

We seem to be staring at the spear point, but what is most prominent in the scene are five skeletons lying in the open, one in the foreground, my knee at its head. He or she lies fully extended, face up. The next nearest figure lies on its side, facing away from us, as if turning away from "the fury of the Saukees" or from us. All the bones remain slightly embedded in the earth, which is brushed smooth and clean. That's what the whisk brooms and paring knives had been for, to reveal our discoveries slowly and carefully. We were to expose what we found but "not to disturb anything."

The site is the Utz Site, on the bluffs four miles south of Miami, the main settlement of the Missouri Indians for over three centuries. Any of those uncovered figures might have been related to the Missouri Princess. The photo dates from the early 1950s, and I accept the ways of the time; but I find it surprising that there was no hesitation about using it for a revised edition in the 1980s and no acknowledgment then of our intrusion since the ethics, and indeed the laws, of excavation had changed. Each of us stares so intensely at the point that we could already be shy of the remains laid out before us. On the other hand, to discover our human past and to wish to inquire into all of it, not as pockets of one ethnicity or another, but as a coat of many weaves and colors—I make no apology for that.

My father and uncle liked to say that they raised no cattle, hogs, sheep, or even chickens, because they wanted winters free for their projects. They had served livestock long enough in their youth, breaking ice in winter troughs. Now they traded income for time and, born hobbyists, became professional. Any research library will offer evidence of second careers that often seemed their first. It all began, I suppose, with their picking up arrowheads as they plowed in the spring on that farm near Auxvasse. But it was accelerated, certainly, by the Spiro Mound.

Spiro, a site in Oklahoma, had been so rich with Indian treasure that it was "mined" in the thirties, in the midst of the depression, with the licensee miners scrounging a living by selling their finds. In their search for salable artifacts, they destroyed most of the mound as an archaeological site. In 1934, headlines heralding "Pharaonic Treasure in the Midwest" appeared in Kansas City papers. The treasure dispersed to collectors and to the curious who would pay for it and so was lost to scholars. Uncle Henry recognized the need for recovering information about Spiro, if not the items themselves. He and Aunt Jean spent years on a salvage operation, tracing the sales and all that dispersal of material as far and as thoroughly as they could. They took up their project while courting, and Uncle Henry published his report in 1952, after sixteen years of research and study. My father contributed 380 photographs, a labor of many winter nights, and Aunt Jean did many of the drawings. This report started their archaeological labors; a second study on the Spiro Mound Coppers appeared twenty-two years later. That time my aunt, uncle, and Eleanor Chapman were coauthors.

On the whole, Unc let the work speak for itself, as he would have had me let Chaucer. He kept the faith of the classical artist that the world is more interesting than he. The mound was there and it offered. Because the mining was messy and unprofessional, salvage was necessary. What more need be said? There may have been stories to tell around his kitchen table, but imposing them on the report lay beyond poor taste.

One sentence in that first report, however, hints at a sweet addition to their story. Stranded within parentheses and distanced by third-person phrasing, it remarks, "The writer first saw the Spiro Mound on the 11th day of June, 1936." I spilled my coffee when I noticed that sentence. "The 11th day of June" was Uncle Henry's birthday. Aunt Jean's was June 6. So for his birthday in 1936, in the midst of the depression, when interstate highways were yet but a dream, Uncle Henry had bestowed a vacation upon himself and had driven from Marshall, Missouri, halfway down along the Arkansas-Oklahoma border to the site of Spiro Mound, on a river terrace on the south bank of the Arkansas River. The trip, down and back, had to cover at least six hundred miles.

June 11 fell on a Thursday that year; apparently Unc treated himself to more than a long weekend. If he first saw the mound on his birthday, he must have set off at least two days before, which is not easily imagined now, given the ease by which we drive that distance and more in a single day. Many of those roads were gravel and the season dust-ridden and dry. That year, 1936, was famous for heat and drought, the worst ever, many an old-timer has said. Those were Dust Bowl days, not just the depression. Unc must have been possessed to have gone to all that trouble and then to have determined it a birthday present he could, at thirty-eight, afford for himself.

Perhaps the present was for Aunt Jean, too, whom he married four years later. They may have set off on her birthday, the Saturday before, and made quite a trip of it together. However that may be, the romance of Spiro had beckoned early on, and my aunt and uncle answered.

Replicas of Spiro Mound ornaments hung on the walls of their living room. Each is a copper sheet, hardly thicker than construction paper and about the size of a legal page torn lengthwise in half. Each has an embossed figure drawn on it with the sharpened tine of a deer antler. Most are hawk warriors, their tail feathers rounding off the bottom of the sheet and a series of vertical, half-moon curves defining their sides. My aunt and uncle teased apart a stack of such plates that had been long fused together. These my father photographed while Aunt Jean made drawings. Then they thought of this more exacting way to better appreciate what they studied. There's nothing like making replicas to help you see the thing itself, they observed. D'Iberville in his journals of French Louisiana (1691–1706) writes of naked young men dancing for hours with flattened copper plates dangling from their belts and making a noise that "assisted in marking the time." Their studies hung like gold in my eyes, and that was just the start.

In their basement, a large, elliptical, Missouri pot with a recurved lip grew under reconstruction. It had been tossed into a convenient pit from which dozens of shards had been salvaged. Father liked to imagine Missouri boys taking the pieces and scattering them among several pits, and "at different levels, just to confound the archaeologists." Native American bows hung on our basement wall, and on a table lay gun parts from a frontier cache. One spring in the thirties, a boy near Malta Bend had turned up a gunsmith's cache while plowing. My father studied the parts and many others from other sites. He learned to restore and shoot muzzle-loaders. He became knowledgeable about the French smoothbore fusil and why, given its relative inaccuracy and the lighter weight of its load, it remained an Indian favorite well into the nineteenth century. He learned to distinguish whether a gun was of French, English, Spanish, or Dutch manufacture and whether from the late seventeenth or early or middle eighteenth century. Such knowledge helped trace the pattern of exploration, colonial influence, and trade on the frontier. His writings contributed to a generation of research on the subject, and he also published a survey of Native American bows.

Few rooms I have ever entered have made me feel more at home than the study my aunt and uncle shared. An oak rolltop desk, a green, wooden swivel chair, worn so that clear wood showed through, an army surplus file cabinet with one drawer half open, two wicker armchairs, a Navajo rug, reproductions of Karl Bodmer paintings on the wall, one of a Missouri warrior, the last Missouris being among the first people Bodmer's party met in 1833 on the Missouri River. Then a bookcase loaded with midwestern history and archaeological reports, manuscripts scattered on the desk and table, on one shelf half a dozen catlinite pipes, my uncle's own pipes in a rack on his desk, a high ceiling, light from a northern window, a framed, aerial photograph of Aunt Jean's homeplace one county west upriver, an 1876 plat map of Saline County, township 52, range 22, that shows the Missouri River sweeping sharply east then zagging north, covering what would become our farm. As that sharp zig-zag withdrew, the river left Tetesaw Lake under the bluff site of the Missouris.

Another underlined passage in my father's *Walden* asks, "What is a course of history or philosophy or poetry . . . compared with the discipline of looking always at what is to be seen?" For Thoreau, the first need was to consider and study exactly what one confronted, rather than overlook it for the sake of more fashionable topics. So first it was the Spiro Mound, "an important part of our history," as my uncle wrote, then the red stone pipes, several bored into broad disks as if they were overturned hats. A pipe smoker himself, he handled them lovingly. Then the pot in the basement, the gun parts my father studied, and the bows. Eventually the green felt of a pool table lay covered with artifacts from a much earlier site, exposed by the 1993 flood. Then, too, partial remains of a butchered mastodon discovered when graders improved a farm road near Miami prompted a quarter century of attention to the implications of that find and its dating. I hadn't yet come upon Thoreau's remark, but among my elders, I found his determination demonstrated.

Jim Duncan Wading

Stately, plump, and mustachioed, Jim Duncan strode into the river, squatted, and watched water close over a belly that looked as if it enshrined a pony keg of beer. Letting both arms rise up on the water, he fingered beneath its surface, working the ground, moving around the edge of a point of land like a blind man hoping to recognize a face.

We were half a section east of the old town of Laynesville, its few streets and wharves covered for decades by the Missouri. In 1993 though, the river, whirling through this bend and taking centrifugal force from the turn, topped and then broke through the levee. It continued swirling and gouged out about ten acres of lake, larger than any farm pond. One result was a loss of more than twice that much from the field, first to the pool then to the new levee that now arcs around it. A graveled road lines the levee top, wide enough in dry weather for two pickups to edge past each other. Duncan's was a red Toyota.

Soon after the flood, George noticed white glitter about a half mile off in a field. Flint pieces lay on the ground as if someone had taken a giant hose and sprayed them in an arc—a couple of bushels of chips, worked blades and points, and thicker, palm-sized pieces edged to make scrapers. Pottery, too, broken into chunks and shards. Soon my father was walking the field with George. They alerted archaeologists, which is what brought Duncan. Signs were of a Black Sand site, named for the stratification of soil in which the first remnant of that culture had been found. If so, this would be an important find, since few Black Sand sites have been recorded. The volume of debris suggested a small town.

The Black Sand people date from before 500 B.C., before Herodotus or Socrates, and so are ancient compared to the Missouri. Their telltale pottery is ridged and decorated. Either they poked a stick repeatedly into the wet clay, making rows of dots, or they made many angular lines, like stacked chevrons. "Incised and incised-over-corded vessels" according to the professional literature. It's said to be the earliest pottery in the Missouri-Mississippi drainage basin. Each bit of information leads to another question. "How," for example, "do we explain as 'first pottery' that which is already decorated?" "Good question," Duncan replied.

Three years passed before trial excavations were made, which brought Duncan into water lapping around his chest while he worked the bank, groping, feeling for more shards or points or chips or pieces of charcoal. Charcoal would be an especially good find since that would suggest settlement. The older levee had covered ground that had long lain undisturbed. Duncan found enough new material to say, "We'll dig here, across this point, southeast to northwest."

"I wouldn't have thought of that," my father said. "He just marches into the water and figures it out. That man knows his stuff."

"He's competent," George added.

After initial excavation, signs of a dwelling appeared. Dark circular patches appeared in the loam where the ends of poles had rotted into the ground. A few of the poles had rotted off and had been replaced with new poles stuck down beside the first ones while they rotted away. The dwelling was only roughly round, like a large ball losing air and caving in on its sides. Its implied diameter was about twelve feet. Was this wigwam a fishing shack from antiquity? It is the first dwelling yet located on a Black Sand site, and it may be one of several that the old levee still covers. Even alone it offers more trace than Laynesville and must have lasted longer if the poles set, then rotted out, and new ones had been placed to rot again.

The 1870s was Laynesville's only good decade. By the middle of the next, it was caving in on itself, even though, in October 1886, the *Miami News*, under the heading, "Local Pride," told of a young girl "not a thousand miles from here," who promptly responded "Laynesville" when asked to name the largest city on the Missouri River.

It would take extensive excavation, the removal of old and new levee by bulldozer, and much more digging to discover the extent of this settlement. The Missouri site overlooking it from the east spreads over three hundred acres, but here only a garden-sized patch had been uncovered, revealing the outline of a single shelter, though the flood had dumped enough material on the adjacent field to promise much more.

The day I visited, Duncan had brought three university students. George was there, as was a high school teacher from nearby, and I had driven from Iowa. Soon we all had our marked-off five-foot squares that we shaved down with trowels six inches at a time. Then we sifted the dirt for what we might have missed. A flint chip or another nubbin of charcoal could turn up in the sifter. Anything found would be recorded as coming from level X of square Y and drawn in on a chart. Most of the richness of the site, however, seemed to have washed across the field during the flood. A later flood may uncover more.

What is the attraction? Why do archaeologists get to dig and sift through sandboxes without end? Why are amateurs eager to help? My father and mother and aunt and uncle had long been members of the Missouri Archaeological Society. My brother and I had joined digs before. Now we had slipped into another on a hot, midsummer afternoon, grateful that Duncan's competent survey had laid out our work in the shade of a cottonwood grove. I caught sight of a jay-sized bird on a lower branch. "A yellow billed cuckoo," remarked George. I couldn't remember having seen one before. The next was a dickcissel, a prairie songbird with chestnut shoulders and a black bib outlined in yellow.

Given my own square, I soon sat in the dirt, which had the patient color of the bottomland I'd often plowed. With the trowel, I shaved off blade-thick slices of sandy loam again and again. It was like picking scabs, or peeling sunburn, or plowing and turning up another and another even furrow. One more and one more, over and over, a physical mantra. Still damp in the shade, and in its earthen form, the loam dried quickly as it curled from the blade and fell. I could pick up a handful and work it apart between my fingers as small fingernail-sized lumps that had suggested charcoal turned out to be sand. Neither gooey nor granular, it worked into dust.

Lives have been given to archaeology and for good enough reason. We seem compelled by beginnings, or at least by ideas of beginning, which is why we treasure our classics. So we get down in the soil and dig back to ancestors who arose from the earth and were reabsorbed by it. Even if the Black Sand people had little to do with "our" beginnings, depending on who is speaking, in a broader sense we recognize a human connection by which our variations are the accidents, not the essentials of the matter. In archaeology, the matter lies in the earth, and whether or not we were made of earth, we have long gone into its making, like the trees decaying in our strip of timber. No doubt someone fishing at this site more than two millennia ago also saw the dickcissel and admired its chestnut shoulder and tinge of yellow. Someone dying lay attended before returning to dust. Though only one true farmer was among us, few of us were many generations from the soil, from one patch of which we now searched for rumors of women and men who had preceded us.

We had no idea how long they lasted or what caused them to float off in their caskets. Perhaps over centuries and pathways too thin to trace they reemerged as the Osage, Ioway, Oto, and Missouri. Perhaps they drifted west or south and evolved into people of a wholly different name. Perhaps, like the Missouris, they died out. Much more remains unknown than known, all of which reinforces our desire to fill in one more piece of an ancient continuum by which gossip modulates to metaphysics.

The Miamis

Henry Ferril and his father, John, had hunted and trapped up and down the Missouri, looking for a town site safe from floods and

*with a landing for boats. He established a ferry in 1836 and laid
out Greenville two years later.*

—*Scrapbook of James Thorp, vol. 5*

In the long history of beginnings, some are newer than others.
Originally Greenville, Miami beribbons the bluff just west of the
state highway that crosses the Missouri at that point. Before the railroads, its population reached 800, some say 1,000; now it is fewer
than 150. Early in the nineteenth century, whites began improvising settlements by following the Missouri upstream, picking out
places on the higher ground. "Greenville" kept faith with a tradition
of naming midwestern towns, recalling the summers that attracted
settlers rather than the winters they endured. Fairfield, Fairville,
Brookfield, Bloomfield, Brookland, Blooming Prairie, Belfield, Pleasant Plain, Belle Plaine, Prairie Rose, Grassy Butte, Greenville, Wild
Rice, What Cheer, Blue Mound, and Blue Earth. Our maps are re-
plete with such names. One sign on Interstate 80 signals "Belle Plaine
/ What Cheer" to passing motorists. But Greenville became Miami
five years after its founding, taking its lasting name from the last
group of native people who had been driven from the place.

Before the Miamis entered the region, the Missouris had walked
those ridges, and before them, the Oneota, the Black Sand people,
and others of whom we know less. Just when the Miamis arrived
is uncertain, but they had been driven west from Ohio in the 1790s,
and by the time of the War of 1812, they were camped in the bottoms northeast of the present town. Like many other tribes on the
frontier—the Winnebagos, Potawatomis, Kickapoo, Sauk and Meskwakie, Ioways, Shawnee, and more—they were in loose confedera-
tion with the British.

Skirmishes were many as were the dangers to new settlers, who
hardly waited for Lewis and Clark to complete their journey before
flooding into the area. Nathan Boone, Daniel's son, led a company
of rangers who did their best to suppress attacks and crises as per-
ceived by the settlers. One of these engaged and removed the Miamis, who must have had the most transient connection with the region of any people we could name. Before that happened though, or
at about the same time, since the records are inexact in such details,
a contrasting story graces our annals.

Jacob Ish, "the first white settler in the area," and his wife and children had driven a wagon into the bottom downstream and east of the present town and had camped for the night. The grass was "taller than a man on horseback, dry and sere," and Ish had let his horses loose to graze in it. Overnight they wandered off, and in the morning he and a son went in search of them. While they were away, "six painted Indians appeared at the wagon, and began talking wildly and gesticulating alarmingly to and at Mrs. Ish."

> She saw a volume of black smoke rising in and approaching from the west. Then the Indians caught up fire-brands, and, setting fire to the grass, soon had quite an area burned over. Into this cleared place they rolled Ish's wagon, and removed all his other property and his family. Hardly was this done when the conflagration was upon them; but the fire passed them by on either side, and there was not even the smell of fire upon their garments.

No smell of fire, perhaps, but surely the whiff of story, and one remarkable for its goodwill. The stories more commonly reported are of settlers, or sometimes their slaves, surprised and slain in their fields or of a brave loosening the chinking in a cabin wall and shooting a man by his fire. Mrs. Ish shared what tobacco she had with her saviors, quite possibly Miamis, and when her husband returned he declared each man a friend for life. Ever after the Indians were welcome in his cabin, and once Ish bought "a whole caddy of tobacco" for them "of which they were very fond."

Around the same time, the Sauk and Meskwakie attempted to repossess themselves of their hunting grounds. Aided by the British, they drove settlers downstream into their forts, drove off much livestock, and plundered cabins "most expertly." Knowing better though than to hang around, they traded their booty to the Miamis, whom they passed on their retreat. Then the Miamis stood accused as the guilty parties with the evidence hot in their camp. Or perhaps the Miamis themselves made a few of those raids. However it happened, at least one stolen rifle turned up in their camp.

Since the Miami, according to frontier rumor, numbered several hundred warriors, the neighborhood militia did the prudent thing and called for government help. Maj. Henry Dodge, who would come into his fame, or infamy, later during the Black Hawk War, when his men slaughtered many retreating Sauk and Meskwakie, brought in a battalion of regulars, placed the militia under his command, and followed the old Osage Trail, soon the Santa Fe Trail, toward the Miamis. On this day, he made a better showing, at least by contemporary standards. His men led their horses quietly through the underbrush, circled the Miami encampment, and attacked. They took the place without a casualty but faced one minor problem—the camp was empty.

Before long, Shawnee scouts located the Miamis in a wooded area a mile or so downriver. Some reports hold they were taken completely by surprise. Others claim that the Miami were fully aware of their danger, had fled their camp to avoid the attack, and were only pretending injured innocence. Most reports also claim they numbered fewer than half their rumored four hundred, and most of those were women and children.

Dodge ordered the property impounded pending an investigation. At that a local militia leader named Cooper, a brother of the man who is said to have been shot while sitting by his fire, and with his eye on a rifle he recognized, drew his sword, laid one hand upon the major, and proclaimed, "By God, Sir, if you attempt to enforce that order, your head will fly off your shoulders like popcorn off a hot shovel."

That threat probably tells us more about popping corn on the frontier than it does about the climax of this story. For other reports have Cooper speaking up but Dodge threatening him with a sword, which puts the opposite head, or breast, in peril. In most accounts Nathan Boone sides with Dodge, even though he too is more or less a local. At least cooler men intervened, and arrangements were made "to the satisfaction of all concerned," as the 1881 *History of Saline County* puts it. But that satisfaction must not have included the Miamis, for the *History* adds, in the very next sentence, "the men from Boone's Lick got their property." The Miami protested but were not believed, and their "removal" soon followed.

One report claims they were returned to Indiana and to the larger part of their tribe. Others say they went west but reserved a right to return to hunt in Saline County, "which they often did." Still others tell of burning "Miami Fort" and "Indian prisoners taken to St. Louis, and exchanged at Prairie Du Chien." That report assumes the War of 1812, in which case the Miamis were not just a tribe in migration under the usual pressures but allies of the English. By the mid-1830s, the cost of "Indian removal" was among the largest items in the federal budget, and soon enough, the Miamis "settled," or were settled, in Oklahoma.

The Old Fort

When Ish and other white settlers moved into Saline County, they followed the river. Large game and small had followed it, as had earlier hunters over a long expanse of time. Clovis points, the earliest known artifact of the Americas, show up here and there in Missouri. The Black Sand people came along about ten thousand years after the Clovis point makers and the Oneota another two thousand years later. Whether Black Sand and Oneota were the same people who evolved through intermediate cultures that only seem distinct to us or whether they followed one another into the region is unknown, though the latter is generally assumed by archaeologists. It is assumed too that the Oneota evolved into the South Siouan tribes of which the Missouri were one, with the line between Oneota and Missouri being the moment of white contact. Here is how Lance Foster, a descendant of the Ioways, first cousins to the Missouris, puts it:

According to tribal tradition, the ancestors of the Ioway Indians united as a people ages ago. The clans had come together and agreed to become a People, the *Honga,* the Great Nation. Some clans had come from the Great Lakes. Others had come from the north, from a land remembered as very cold. Yet others had come from the western prairies or the eastern woodlands. Some of the ancestors had made great mounds in the shapes of animals and birds along the bluffs of the Great River. Others had traded down the river to the great southern mound cities and returned with new ceremonies, new beliefs to add to the older ones.

This development of the clans into one Nation is traced in the ancient stories and traditions of the Ioway and their brothers: the Otoe, Winnebago, and Missouria. Other relatives of these peoples also seem to have been part of this Great Nation, including the Omaha, Ponca, Kansa, Quapaw, and Osage. Stories recall a time when they were all one People and when all the land of this Middle Place was theirs. These stories seem supported by archaeological research. The culture is called Oneota, after a rock formation along the Upper Iowa River in northeastern Iowa where certain types of pottery fragments that characterize this culture were first found.

Foster's account explains more or less the linguistic affinity of these peoples. His underlying assumption of great lengths of time could explain how people coming from the Great Lakes met others from the Ohio and Tennessee valleys and they became a Great Nation, sharing variations of a language. Much remains mysterious, but the Oneota must connect with the mound builders, for they moved a lot of earth.

They made mattocks by binding the shoulder blades of bison to wooden handles, and they dug and dug. They dug trenches, six feet deep and three to four feet wide. Today we surmise the Old Fort from two roughly parallel ditches that outline an area larger than two football fields. Before white settlers moved in, that treeless prospect, high on the bluff, would have afforded a panorama of the floodplain, the river, and Bottomless Spring. The site blends with that of the later Missouri town. The princess would have been standing at its edge at her homecoming, had she made it all the way home. On a clear day, she could have seen a glimmer of Fort Orleans in the distance to the west.

The "Old Fort" may not have been a fort. With no internal supply of water, it could not have withstood a long siege. It has also been read as a ceremonial enclosure. Still the Oneota had pottery, the spring, and a river at the base of their bluff. Perhaps they stored water with the same discipline that they dug trenches. Perhaps they conceived Hope as huge furrows and so performed a ritual plowing, a kind of prayer. The more they carved those bluff tops, the more their corn would flourish.

However that may be, the Old Fort is a magnet for stories. The 1881 *History* claims abundant finds of pikes, hatchets, axes, clubs, and points, both stone and iron. "An iron crown was also found here, indicating that, somewhere in the far shadowy past, royalty dwelt in these fair and favored regions." The writer suggests a race of giants. Another interpretation considers the trenches the remains of Fort Orleans. Some early writers had difficulty crediting indigenous people with its construction. Others though have speculated that its builders were ancestors of the Aztecs. Contemporary archaeology seems satisfied to know the work as Missouri-Oneota. But it is difficult not to lean toward the *History*'s conclusion "that the whole of this great valley was once, in the far distant past, the empire of a vast population" and to note how that surmise parallels Foster's synthesis of Ioway tradition.

The Oneota-Missourias chose their site well as "the people who arrive at the mouth." The Osage Trail must first have been theirs, a natural trace under the bend of the river. The squared-off section lines our roads and fields follow would have been anathema to them. By the time settlers arrived, the Missouris were but scattered remnants. Still, from their bluff site, they had once looked in all directions and called as far as they could see "home."

One hundred generations, give or take a little, connect the Black Sand people and ourselves. White contact with the Missouris occurred about nine-tenths of the way along that path. Even with the first generation back in my family, I am reduced to fragments and traces of story. Did the Missouri Princess really stay in Illinois and never once return home? When Bourgmont returned from his first visit to France, Missouri warriors came downriver to meet him near where St. Louis now stands. For the Missouri party, a two hundred fifty–mile trip was no extraordinary expedition. Did it never happen another time, for the sake of the woman?

Or what of Petit Missouri, whose life takes the shape of the Missouris' vanishing? From Paris to the deep plains, his life crossed numerous borders. Just as Bourgmont prefigures Lewis and Clark, Petit Missouri prefigures Sacajawea's son Pompey. Clark called him "my boy Pomp," sent him to England and oversaw his education. Pomp returned to the frontier as a mountain man and was last reported mining for gold in California. Petit Missouri fades more completely from our picture. For a time, though, he must have wandered the same bluffs and known the Old Fort well. That, at least, is where I have often imagined him.

Perhaps he carried a French fusil; let's call it his inheritance. The bluffs provided a vantage over the wooded bottoms. Wandering buffalo on a sandbar could propel hunters as quickly as the wind changes direction.

Petit held his weapon high, pulling it past grapevine and underbrush. Down in the timber, he and some companions ran quickly and almost noiselessly along a game trail, leaping over logs, dodging low branches. If they didn't hurry, the bison might have moved on. If they rushed recklessly, they would betray their approach. If they veered off course, they might come through the timber half a quarter away. A party of Sauks might have seen the same few cows and bull and be on their way by another path, with this hunt soon to flare into a skirmish.

They ran through the heart of the heart of Missouri country, which could only increase their confidence. Willows grow closest to the water, and a stand of them was a signal to slow down. Picking their way, they paused behind the last leafy screen to load the fusil. Had Petit loaded it earlier the shot could have fallen from the muzzle while he ran. He had to be careful not to let sweat dampen the powder as he poured it in. A cuckoo sang, but the bison failed to take that as warning. His weapon loaded, Petit and his companions picked the closest cow and found angles of attack. Three of them drew bows, and all four shot together. A cow fell at the water's edge while the bull and three other cows splashed past the hunters, clambered up the sandy, crumbling bank, and crashed through the brush.

Now their work began in earnest, skinning, butchering, and loading the meat in slings improvised from willow shoots and bark. Even shouldering all they could, they had to bury much of it in wet sand, wrapped in the cow's hide, and hope they could make it back before wolves beat them to their cache. They hiked several miles to their town and returned as dusk thickened. On their second trip, carrying everything they could salvage, a little drunk on weariness and success and moving more slowly, the men began to place themselves in stories told of hunters for generations.

They tried to imagine how many parties like their own had run down off those bluffs to surprise a bison, or larger game as their oldest stories hinted. They guessed that the number of all their steps homeward would not be too many. No matter their successes, the path they took was into an abyss more bottomless than the spring. Petit Missouri, whose story I know so little of but who figures the Missouris for me, left no physical trace. As far as we are concerned, had his name not appeared in a few French texts, there would have been "no boy at all."

IV
Mother, Father, Farm

The Miami Mastodon

*What do we want most to dwell near to? . . . to the perennial source
of our life, whence in all our experience we have found that to issue,
as the willow stands near the water and sends out its roots in that
direction.*

—Last underlined passage in my father's Walden

Tchuk, whap whap; tchuk, whap whap; tic tic, whap, whap whap;
tchuk, whap whap.

As I remember it, we spent the whole day driving finishing nails
into plywood subflooring. Thousands of nails, and many more than
two-thirds of them driven by Junior and Sam, true carpenters, one
white, one black. I was their summer-long assistant. From them I
learned to stand in the wispy shade of the locust on the hottest
days, one of which reached 116 degrees Fahrenheit, and let its long
seedpods act as wind chimes. We drank black coffee at the lunch
counter out on the highway, letting the coffee warm our stomachs
so, Junior said, we would feel the sun's heat less. I still don't believe
that one and prefer cool water in the shade, but I was a college
boy who could stand most anything, and hot coffee was Junior's
summer habit.

The plywood covered a rectangular basement of cement walls
that we had already poured and finished and that held casement
windows in the north wall, which looked out over the river from the
loess bluff at Miami. After thirteen years of farming in the bottoms,
my father and mother were building the house on the hill they had
dreamed of, and as Father had promised Mother when we moved

from Chicago. But the right site hadn't been easily found, and much else had had priority—clearing land and starting a farm to begin with. Now though, as I was close to finishing college and George was about to begin, the Miami bluff caught their attention.

I didn't know it at the time, but with a baseball, and luck, I could have hit Mart Rider's sagging gravestone in an abandoned cemetery across the road. Some mastodon bones would turn up thirteen years later less than a half mile east of where I stood. About that far west was where, in Mertens Hall, Blind Boone had offered "by far the best entertainment given in Miami for many years." The sun had swung well into the west, and the long shadows of trees had begun to crosshatch our deck. Tchuk, whap whap. I tried all day to match that rhythm. Junior and Sam would each set their nails with one tap and drive them home with two strokes more. While they drove those nails firmly into the flooring, their left hands would be singling the next nails from the handfuls they carried and placing them for the next setting taps. They swung their hammers easily from the butt ends and never hit their thumbs.

I could wield the hammer too without choking up. But to set the nail with one tap and drive it in with two strokes more took practice and more confidence than I had about my thumb. So I was a tic tic, whap, whap whapper and sometimes whap whap, whap whap. Every once in a while, as if by accident, I found the rhythm— tchuk, whap whap. Both Junior and Sam would catch the altered beat, look up and smile. Give me a little time and I could approach competence.

Finally we got off our knees, having no more nails to pound. Father had arrived from the farm with a fresh jug of water. Sam and I ambled off into the shade of the locust to refresh ourselves. But Junior's thirst was different. Junior was our head man, our carpenter-in-chief. Junior took just one sip then got his level from his truck and walked back out on the floor. He set it down and checked, picked it up, walked a few steps, set it down and checked again. Over and over he picked out a new spot on the floor and set his spirit level down. The bubble balanced every time like the checkbook of a very prudent man. That was all the refreshment Junior needed. Each time he bent over, he stood up more cooled off. "Ain't that perty," he murmured, "ain't that perty."

It took another thirty years before my father interviewed Paul
Stonner. By then the two men were among Miami's oldest citizens.
My father had long since stepped back from a dozen years as the
town's mayor, but even so he was a newcomer compared to Stonner,
who had lived in and around Miami since 1920. The old Miami that
Stonner remembered had two hotels, six stores, two garages, two
barbershops, two drugstores, four doctors, one lawyer, one dentist,
one newspaper, one bank, four churches, a high school, a grade
school, and a livery stable. It had a town band and a hitching rack
that led around one corner off High Street and all the way down the
hill. Some days there'd be a quarter of a mile of horses, buggies, and
wagons hitched up and waiting. Soon they were talking about the
1951 flood. Stonner first:

> "I saw slabs of concrete perty near as wide as this room and as long
> as this house that washed up way down there in the field. Took the
> railroad track out twice. You could go out on the bridge there, and
> from the warehouses, you know. You could see halves of beeves by the
> hundreds. . . ."
> "Floating down the river."
> " . . . floating down, dead hogs, dead horses and cattle, what looked
> like complete houses. Every now and then it'd look like a complete
> house. The roof was setting up there, sticking out of the water, and
> you'd see part of the house, skating right on down to New Orleans."
> "Well, they told me that you were the man who reported that you
> saw a house floating down the river and a woman in the back yard
> hanging up the clothes."
> " . . . Not quite."

My father gets Stonner to hesitate, for an instant. Then he continues
with a tale of brushfires burning so hot, back when they cleared land,
that it turned black gumbo red. Countering all that, though it is not
on that tape, my father could have told Stonner a story every bit as
tall, many times as deep, and more remarkable still, true as far as
we know.

In 1973 graders cut into the bluff at Miami to improve a county
road. Checking on what was being disturbed, my father saw a white
patch gleaming from the fresh-cut soil. The find proved to be mastodon bones, ribs and tusks, stacked and fused, with a few tools
beneath them, two pieces of limestone the size of brickbats, which
had probably been used to break up the bones, and three blades, or

scrapers. Carl Chapman, the investigating archaeologist, died before he fully examined the material, which had by then languished in storage for nearly a decade. Some pieces were lost in a fire, and no one went to work on the rest. My father finally reclaimed what he could when a scientist at Woods Hole, Massachusetts, a collaborator on another project, offered to run a carbon-14 test on a sample. The improbable word came back, "35,000 years before the present, plus or minus 1000 years." "There has to be a mistake," the writer added, but a second test only corroborated the first.

Tests though are subject to challenge, and there was reason to doubt this test on this kind of material. More time passed before a researcher on the opposite coast ran a radically different test and came up with a tentative date of 41,000 years ago, plus or minus 6,000. Science places the earliest people in the Americas as just post–Ice Age, only 12,000 years in the past. Suddenly the plus or minus variable is half as long as the whole range of time that had been considered possible. The new suggestion dwarfs the interval between ourselves and the Missouris, or between either of us and the Black Sand people.

Researchers today hope to determine the age of the loess soil in which the bones were embedded and its history as a windblown surface. Bone could not blow in on the wind, nor would stacked ribs have been placed there except by people who butchered their kill. So far new tests "support the antiquity of the Miami Mastodon site," now thought of again, though still tentatively, as 35,000 years before the present.

One doesn't have to be a specialist to know that the long-held date of arrival as 12,000 years ago is open to question. Recently, a South American site has been accepted as 13,000 years old, and those people would have required considerable time to migrate south from an Alaskan point of entry. Several other widely scattered sites are hard to reconcile with the prevailing story. Native American traditions, in turn, have long favored a much longer history of themselves, and though not science, neither are they necessarily incorrect.

My father believes the grader dug into a cache left by hunters more ancient than any of whom we know, hunters who made a kill similar to the one I imagine made by Petit Missouri and his companions. At Miami, the bottoms narrow to under three miles from bluff to bluff whereas both upriver and down they swell again to more than three times that width. Miami commands a bottleneck, an advantageous site from which to watch for big game. A mastodon dwarfs a buffalo. Early hunters would have had no choice but to store most of that kill in the dirt. Then misfortune might have prevented them from returning to their store. Or one of the storms that blew the loess bluffs into being erased all trace of their deposit, and the bones waited millennia for the bulldozer.

The earth doesn't take long to absorb most human trace. While it seems self-evident that bluffs along a major river would attract human attention, we rarely suspect the stories we walk over. Perhaps wanderers migrated between ice ages, arriving after one and surviving another, having learned to live along the ice. Perhaps men and women walked Tetesaw Plain and Bottom for vast lengths of time satisfied, or constrained, to leave hardly a mark. Peering over treeless ridges, they too surveyed the valley through which the river flooded then meandered. Mastodons were the whales of their weather eyes, as bison would be for their descendants.

Ice

We say "river" half forgetting that we mean "bank," French *rive, ripa* in Latin. Bank dwellers, we define "river" by its enclosure rather than as the "fluence" that flows or floods past us. A fish might have done it differently, and so can winter, which arrives *("a-rive")* at riverbanks with a vengeance.

In December 1837, a year before Miami had been laid out, but by which time settlers were bunching up on both sides of the river, a Reverend Eli Guthrie of DeWitt, on the north bank, ran a ferry that joined the two places. John McMahon and Perry Harris operated it for him and both lived on the Miami side. McMahon was the son of one of the first settlers in the Miami bottom; Harris, who was only nineteen, boarded with McMahon and his young wife and child. On this December evening, chunks of ice floated on the water, crowding at the bends, and the river was beginning to "gorge" at some unseen point downstream. The two young men were determined to cross and set out in a flatboat large enough to carry a team of horses.

The night was cold and growing colder. In the early dark, ice chunks already seemed larger on the water. Guthrie urged the men to spend the night with him, but McMahon and Harris insisted on crossing. Guthrie followed them to the water's edge, trying to change their minds, but whether a young wife was calling or a girl-friend or both; or whether it was the bravado of being young and as bold as ice themselves, they refused to listen. Within yards of the shore, ice thudded against their boat. Their poles and oars became straws against the river's force, and the ice and current swept them away. Guthrie and a few neighbors ran along the bank, trying to keep abreast of disaster. They saw the boat ram against a sawyer and McMahon and Harris leap off.

A sawyer is a whole tree, or most of one, washed out of the bank and carried downstream until its root ball or cracked trunk lodges against something and sticks in the river as if rooted. It sticks up out of the current, branches extended, like a boulder in white water. This one was boulder enough for the flatboat and upended it. Poles and oars fell away, the boat rushed downstream bottom up, and McMahon and Harris jumped onto the sawyer's branches and clung there, shouting for help.

Guthrie ran back to DeWitt, untied his skiff, and with William Smith and Lilburn Barns, ventured a rescue in the dark. The time for caution had passed, or at least these men suppressed it. They made it to the sawyer, where McMahon, grabbing the skiff's chain and pulling it to him like a puppy jerking on a leash, drew the bow right up on the trunk, swamping the stern and spilling Guthrie and Smith. Both men went under in a gulp and were never seen again. The boat swept away, and at the lurch, Barns, who sat forward, either leapt or was launched onto a floating piece of ice, his momentum carried him to another, and he kept on running and jumping from cake to cake until he got within a few feet of shore, where he fell into the river and waded out.

Marrying need to the moment, Barns improvised luckily. When he dragged himself out on the bank, he was a hundred yards down-stream. His companions were lost, and the men they had hoped to save still clung to the sawyer in the night.

News of the disaster spread faster than winter dark. First on the DeWitt side and then soon enough where Miami was coming into being on the south, people built bonfires of driftwood to cheer their vigil and to sustain Harris and McMahon, as best they could, with at least the thought of warmth. No one else dared launch a boat that night. But they shouted encouragement and stared as if thought alone, or the penetration of sight, could tie those at the fireside to the very nearly lost. By daybreak, McMahon was dying. Harris shouted the news to shore and with it McMahon's assurance that he wished all his friends to know that he died "resigned to the will of Him who doeth all things well." Not long after, Harris called out that his friend was dead. He wrestled the body higher in the crotch of limbs and secured it to the sawyer. McMahon's relatives urged him to strip the body and to use the clothing to better protect himself, which Harris also managed.

So passed one bitter winter Tuesday night in the long history of the Missouri River. On Wednesday afternoon, men from Miami loosed the ferry on their side and tried to rescue Harris. They didn't get far before they too struck a sawyer. Their boat shipped water and ice, and the men barely managed to scramble back. Other boats were launched from both sides with men now here, now there, thinking they saw an opening and could brave it. But one by one each faltered and turned back. A chunk of ice would strike a bow and ram it off course. Another might crash amidships and threaten to board like a blind and clumsy swimmer. Surely the knowledge of one failure after another weakened succeeding attempts. If Guthrie, Smith, and Barns had been winged by little thought, anyone who tried now was burdened by too much of it. Someone located a lump of lead, tied a line to it, and man after man tried to heave it to Harris, but he was moored beyond their best throws. Like the river itself, rescue was icing up.

Now the only hope for Harris was to endure until the river "gorged" and he could walk off over ice. Other ideas must have tempted him. He had seen Barns hop from cake to cake to safety. Clinging to his sawyer, Harris must have thought and thought about when to leap, which moment to try, which string nearing was sufficient. But initiative ebbed until what little remained froze over. Nor was he launched by accident, flung beyond the paralyzing cast of thought.

Another possibility would have been just to hop a cake and chance it. Perhaps it would run into the bank or onto others that had. Then again, perhaps he could wait long enough.

Meanwhile his family, neighbors, and friends became as rooted to the shore as he was to the sawyer. They could neither assist Harris nor abandon their vigil. Death by exposure was rooted in their heritage. Sam Irvine tells of the earliest days of his family in the county, of their living "out on the prairie," exposed to the winds, and of townspeople believing no one so situated could survive the blizzards. A day or so after each winter storm, a rider would venture from town to find out whether the Irvines were still alive. But the Irvines had a cabin and a fire.

For three days and four nights Harris clung to the sawyer, hoping, waiting, frozen, freezing. That last night the ice chunks began to wedge together, gorge up against whatever braced them farther downstream, and it seemed that rescuers might get to Harris. But Old Man River, or Mother Nature, blunted that last hope almost as quickly as vision of it formed. Two slabs of ice swirled together against the sawyer, "chunked up," and crushed him. McMahon's body was thrown into the river and lost like Smith's and Guthrie's. Harris's stayed wedged to the sawyer so his neighbors finally got it out.

Chicago Women

By rare coincidence, like tulips shining after a spring snowstorm, twice in my family Chicago women married Missourians and made homes in a place strange to them. The oldest sister of one had a long career as a painter, and she too painted capably. The father of the other trained in architecture but preferred painting and became a director of fine arts instruction in the Chicago Public Schools. The first was my grandmother, she who left her false teeth in a glass on the corner of her desk, who played Chinese checkers with me and gave me German lessons, who instructed—or someone very like her did—Blind Boone. The second was my mother.

The oil paintings on our walls may have been prized mostly within our family, but all were originals. A bank of my grandfather's landscapes lined one wall. He died in his forties, and I never knew him. He loved to paint the dunes on the shores of Lake Michigan, their sandbanked gullies, the washes of sun contrasting with a fringe of trees, a bay cutting into the foreground or a broad expanse of lake shining in the distance, all shot through with purple, pink, and bronze as if he had taken hints from Monet. Mother's favorite, I believe, and certainly mine, imagines a mysterious forest pool, wreathed in overgrowth and shadows through which black contests with purple.

We had several of my great-aunt's still lifes, in rich, dark colors with light reflecting off grapes or glinting in the glassware. One of my earliest memories is of astonishment at seeing light emerge just from canvas on a shaded wall and to realize that paint, not glass, not direct sun or electricity, had worked that magic.

One of her paintings centers on a graceful, elliptical china vase holding two long-stemmed lavender tulips. It has long been placed above a small chest on which that vase stood. So I experienced the double take of the art object and the painting of it, and then the conundrum of which was the "real thing," when I discovered that by adding a tulip, Tante Magda, as we called her, had improved upon the vase's display of flowers.

A reddish brown portrait in iron oxide, or sanguine, shows my great-grandfather in profile, wearing a jaunty, wide-brimmed, almost western hat with a full beard balancing its backward sweep and crown. He had fled European revolutions in 1848 and had got as far as Chicago, where he set up as an apothecary. Eventually, we found in the attic a portfolio of Tante Magda's studio sketches, in charcoal on paper, done in Munich in the 1880s. We framed and hung the ones not too creased and torn. One shows a sad-eyed older man with bald crown but locks over his collar, then down the right margin of the page, four other attempts to render those downcast eyes and furrowed brow, the most finished of which peers over the man's left shoulder. Though only a sketch making efficient use of paper, it seems a modernist study juggling several perspectives.

My favorite is a nearly life-sized, full-length, nude portrait of a young boy. His hands hang straight and loose at his sides. He looks slightly off to his right. A coin on a string rests just above the separation of his chest, dropping toward his belly button and penis.

Sadly, this portrait had been folded over for so many years that it was all but torn through at the boy's waist, and pieces of paper had fallen away. So we sacrificed the lower part then framed and hung the upper half. When my daughter was close to that age, visitors often asked whether she had been the subject. Later they asked the same about her younger brother.

The single painting that I remember most is by my grandmother, though she was less the painter. Simply framed in quarter-inch round, it depicts a path ascending in woods. Trees climb the left side of the canvas and curl over the top; a hint of meadow breaks to the right, with light pouring in from that side and hints of blue sky showing through the trees. The rising path curves toward the upper center of the canvas and disappears above eye level. It was a landscape more nearly my own, and a road more taken. It hung in my bedroom for years.

I do not know at what age Anna Heuermann left Chicago for Kentucky, but she was twenty-eight when she married in 1896, having taught in three colleges by that time and having lived in Missouri for six years. She went first to a position that had been located and agreed upon through the mail; and my grandmother, when prepared to leave, went downstairs to find Magda, her oldest sister, dressed and seated beside a suitcase of her own at the front door. Their mother had been dead since Anna's early childhood, and Magda, the veteran already of European study and travel, was taking charge of her baby sister, whom she intended to make sure was suitably placed.

Those two outlived their three intermediate sisters and were reunited a few doors down the street from us in their final years. My father says they fought all the time, which I do not remember. But when I entered their home, I usually found them in separate rooms. All of which must have started early, long before that moment by the door when Anna looked down hard on Magda and said, "If you stand up, I will sit down; and I will not budge from here until you have seated yourself." No doubt she delivered her ultimatum in German. That once at least she got the best of her sister.

"I am five feet," Tante Magda once said, "and the only thing I was ever vain about was, what shall I say, the longness of my hair and the smallness of my figure. My hair used to touch the floor, with six inches to spare. When I was a student in Munich I wore a fedora. You know what a fedora is? A fedora, and a tailored dress with a high, stiff collar. I did think I was perfectly wonderful then. I was eighteen when I first went to Germany and had the handsome Franz von Stuck as my master. Ah, München, München."

She had passed eighty by the time I was born, and it was most of another decade before I saw much of her. A rosy-cheeked putti looked down from high on her wall, the only glimmering thing in a dark corner of heavily upholstered wooden furniture. Tante Magda pushed her walker toward the kitchen to pour me a glass of milk. "Would you let the Venetian Blinds come into your home?" she asked, nodding at her curtains as we left the front room. I had no idea why she asked or of what exactly she spoke. I guessed she scoffed at some charitable order.

Following her into the kitchen, I passed a half gallon tin can crammed with paint brushes, all sizes of tips, blond tips, clean and shaped, a bouquet of brushes to which she resorted daily. She was held together, I imagined, by sixteen-penny nails. "Pins" is what I had heard, for both hips had been broken and her back cracked since she had turned eighty. Cataracts curtained her eyes, and perhaps that is why the still lifes of the forties, as she approached ninety, are increasingly dark. Perhaps they became her own ironic version of the Impressionists.

Once when being presented to Kaiser Wilhelm II, who sat for a portrait, Tante Magda touched, indeed stroked, his dress sword, which seemed a scandal to the company, but she got away with it as a young *"fresche Americaner."* She was still fresh years later as company assembled for my parents' wedding picture. The photograph shows her bending at the waist, from the far right side of the front row, her hands clasped behind her as she makes a right angle and tries to obscure three of her sisters.

What effect, I wonder, did she have on my uncle and father, who grew up hearing often of their colorful artist aunt. They made summer trips to Oak Park from their Missouri farm and saw suggestions of a world pitched far beyond them. Her still lifes may be dark, but light dances in the crystal. Her paintings proved windows with views. When, in his early twenties, my uncle served with Quaker relief organizations in Poland, Tante Magda had ventured there already. Her mother had a Polish name, Sabransky. The Tanten spoke German at home. Uncle Henry communicated through German with his Polish and Russian colleagues.

Or there's the story of Tante Magda on the Erie Canal, taking a boat to Utica, and her boat being run down by another. I have no idea why Utica or what was the occasion. "My dress was caught between the two boats and I went down into the filthy water—way down into it—in which I had seen carcasses of cats caught in the locks! Drowning did not seem to cause anguish, but the filth did. In the nick of time one of the many, many onlookers"—Was there always an audience?—"threw off his coat, jumped into the water, and I was saved! At Utica, after a bath and fresh, clean, clothes, I sat in the sun surrounded by newspaper men and staring crowds, and heard the words, 'Beautiful mermaid.' My hair was black and fell like a silken mantel to my knees. The effects of my bout with Neptune did not wear off for a long while."

She had watched Union Troops march in Chicago and had witnessed the Chicago Fire. A clipping from the *Chicago Herald* (January 23, 1917) features her on the same page as Whitman, and a piece of the gown she wore that day is pinned next to the newspaper page in her scrapbook. She won a gold medal at the Columbian Exposition of 1893. In an essay on Herr von Lenbach, one of her portrait subjects and another master in Munich, she wrote of art as being wholly concerned with locating "the infinite embodied in the finite." Her specialty, until her eyes failed her, were portraits in miniature.

To her bungalow on Eastwood, Tante Magda brought a rosewood spinet piano. She stood on tiptoe to light the candles on it. A small Persian carpet lay atop it, and a larger one spread across the floor. A dark, high-back couch, her still lifes wherever there was wall space, and the putti overlooking us from high in the corner completed her decor, except for the fourteen-inch plaster of Paris mock-up for a statue of a seated Dante that she traded a painting for with a fellow artist in Italy. It now oversees my own desk work, eye to eye with the kachina.

Some Sunday afternoons the neighbor preacher came to call, the one with whom I sat watching Friday night fights. He had put away his cutaway and donned a lighter linen suit. "This spinet, Dr. Smith, is the only thing in the house that doesn't work," Tante Magda purred. "It cannot make a sound. When are you going to take me dancing?" Her window opened on a universe other than our rivers and fields, and Grandmother, apparently, adopted much of Tante Magda's breadth of view.

But she went to Kentucky, rather than to Munich, then later to Missouri. "At the time," one study of Blind Boone reports, "the only songs he knew were cheap songs he had heard and learned to play on the streets. Then a music teacher at Christian College took an interest in him and taught him to play some classical songs." A parallel account places this instruction "in the early 90s," by which time Grandmother was professor of piano at that college and a few years younger than Boone if he was her student.

By the time Grandmother came to Missouri, she had become a horsewoman. I have no explanation for this except to note that she had gone to Kentucky first and must have been eager, a city girl and a German immigrant's daughter, to take up the sport of the gentry in her new neighborhood. So she arrived in Missouri sidesaddle, you could almost say, her long skirts flying.

My grandfather had grown up with horses, but his horses pulled buggies and plows. He played the guitar and formed, with his brothers, a "Violin, Mandolin and Guitar Club." An old photograph shows the quartet—a cellist having joined the group—playing beside a fine brick building with Corinthian columns. Two young women have stepped out of an open door onto a balcony overhead. One of those women may be Grandmother Anna. If so, Grandfather keeps his back turned and faces his fellow musicians rather than her. The young men were town, not gown, and did not themselves go to college.

Music likely brought my grandfather and grandmother together. My father tells of once bringing his guitar home from college. "Father picked it up, tuned it differently from the way I had been shown, sat down and played for about an hour. He did not chord as I did, but *played*; melody, accompaniment and everything." Then he put the guitar down and could never be coaxed to it again, which mystified my father. Grandmother, in contrast, never stopped being a music teacher, though she became a farmwife. She gave private lessons throughout her life, published books of piano instruction, and edited a journal for a national society for music education.

Twice Grandmother designed houses that she and Grandfather built. Desiring running water, she designed both homes to hold a thousand-gallon tank in a second-floor room especially constructed with extra framing to give the tank sufficient support. It caught rainwater off the roof and ran it downstairs to the sink so that she had soft water to lavish on her hair. Obviously Grandmother and Grandfather also thought it worth their while.

Grandmother painted their wedding portrait in shades of blue on a tile plaque. Floral columns frame the couple, with their initials and wedding date worked in. Grandmother's hair is up in a coil, and the clean line of her neck falls inside a cotton ruffle. She's younger than my daughter. Grandfather's white collar and the left shoulder of his dark suit show. In 1927 a tornado struck their home, scattering it to kindling. Poking around in the wreckage, my uncle found pieces of this plaque, which he glued together and filled in with plaster so Grandmother could repaint the missing parts. Until recently, it hung in the house that I helped build and that Mother designed to exploit wide views of the river from both the basement and the main floor. Now it hangs in my home, for rivers and water alone are not "the perennial source of life."

The Bridge

One October Saturday, George drove me into the bottoms and dropped me off at one end of the levee to walk out at the other. Corn had ripened to gold but was not yet dry enough to harvest. Willows and cottonwoods along the riverbank wore autumn yellows. The first small reconnaissance flights of ducks and geese hinted at ancient migrations about to begin. An occasional freight train passed across the river to the north and transcontinental jets overhead, but I walked a dozen miles of levee without seeing another person. Neither hunting nor harvesting had begun.

It had rained an inch and a third the night before. Under blue sky and a scattering of clouds, I wore a windbreaker. In my notebook I wrote,

Poison ivy, willow, cottonwood, wild plum, sycamore, soft maple, mulberry. Mulberry and maple fill in below the cottonwoods. Smartweed, milkweed, sunflower, grape. Dock, lambs quarter, locust, box elder, water lilies. At a distance, I thought the lilies were pelicans.

Buzzards, monarchs, dragon flies, smaller butterflies with orange tipped wings. Kingfisher, blackbirds, cormorants, a flock of them, heron, goldfinch, a Cooper's hawk—beat, beat, beat, glide. Blue jays, grasshoppers, crickets, deer tracks, raccoon tracks, their distinct toe nails, brown woolies, a redtailed hawk, crow, pelican, deer. Two sharp shinned hawks flare up from my right, lifting over the levee toward the river. Doves, killdeer, flicker, tree swallows, four wild turkeys, then a flock of them following. Redwing blackbirds, barn swallow, mallards, marshhawk.

On the island I surprised two does, windward. I came up over a rise on an old levee and there they were, staring at me.

Black plastic scrap, shotgun shells, a circle of bottles laid out as if to ring a fire.

From that ring I could see the bridge at Miami.

The bridge came late to Miami. On other roads, upriver and down, bridges crossed the river. But from the time Henry Ferril established his ferry in 1836 until 1940, ferries were the only way across at Miami. Turn of the century papers tell of Miami and DeWitt playing each other, one from each side of the river, and that was an event, getting a team and supporters across and back for the afternoon. Each town had a cornet band, for crossing merited a party.

Once too there had been a one-room school in the bottoms across the river. A one-room school meant one teacher. Normally, when a teacher came to live in the area and teach in such a school, neighborhood families took turns providing room and board. For some reason, this did not happen with that teacher in that district. She was undaunted, however, and unwilling to lose her position. She took a room in Miami and kept a rowboat tied up at the bank by which she rowed herself back and forth. When the river iced over, she walked across, carrying a plank to lay over soft spots.

Early ferries were strung on rope, then later on cable. A horse wound the rope around a winch, trudging in a close circle on the deck. Then the railroads came through, both to the north and to the south, and cut Miami off. After a while there were fast mail trains, but they wouldn't stop within a hundred miles. A local judge called the railroad and asked, "Would you stop for twelve men." Yes, an official said, they'd stop for twelve men. So the train stopped at DeWitt, and the judge boarded. "Where are your twelve men?" inquired the conductor. "I didn't say I *had* twelve men," the judge replied.

For years, Miami citizens were interested in a railroad spur, connecting them north and south through Marshall to Sedalia, and in one account all the way north to Des Moines. A 1901 notice mentions needing subscriptions from Miami, that Marshall and Sedalia "had done their part." That would have put Miami back in the action, but the project came to naught.

Meanwhile steamboats and barges docking at Miami needed their goods transported inland. Wagons and teams of mules often took two or three days to cover the fourteen miles between Miami and Marshall. Wagons had to be hauled up out of mud-filled washes. Muddy Creek blocked the way, then farther south, Salt Fork. Men pried the wheels up out of mud and onto brush cut and laid for better traction. They slept on wet ground beside the wagon. "The good old days; piss on 'em" is how Paul Stonner summed it up.

But toward the end of the depression with America gearing up for war, the idea of a bridge found its moment. Workers came, public money abetted the project. Building the bridge employed 150 to 200 men at a minimum wage of forty-five cents per hour. The work went on for most of three years. "Miami now has a new hotel, a cafe, beer tavern, new filling station and its first night club," claimed a story of April 10, 1939. Miami thrived. Senator Truman attended the dedication in 1940. Speeches were made and ribbons cut. The total estimated cost was $534,007. That last seven bucks must have gone for ribbon. The toll was thirty cents per car plus a nickel for each passenger.

My Aunt Jean brightened like a butterfly when I asked about the bridge. She remembered a day when a beau took her to see it. They crossed the Missouri at Lexington, drove downriver on the north side, crossed again at Miami, had dinner, and drove back home to Lexington following the south side of the river. "Oh my," she said, "two crossings in one day; that was something."

From the first, papers heralded the bridge. Here was a chance for Miami to regain ground. Forty Miami businesses had advertised in an 1874 publication. Miami then had two hotels. It had a savings bank with $50,000 in capital secured in real estate. There was a steam sawmill, a carriage factory, a flour mill, and a steam ferry to and from St. Louis. Stage lines made daily trips to Marshall. During 1873 over one thousand carloads of stock crossed at Miami. Railroads paid the ferriage and would receive and deliver in Miami. But by 1940, its population had dwindled to two hundred, and the Marshall paper allowed as how, with the bridge, Miami could hope for renewed prominence. What was good for Miami, it added, should be good for the county.

But my aunt wouldn't have had nearly the afternoon she did if Miami had been the destination rather than a point of passage. And a little town justified a bridge much less than did the highway beside it. A recently paved road ran straight into Marshall, which had a square with businesses all around. Moreover, the Marshall merchants had a plan of their own. They brought out a coupon with seventeen tabs, each one good for three cents off any dollar purchase. Fifty-one cents bought an excellent lunch. Or it almost reimbursed a round trip across the bridge. "Thus," summarized a Miami paper, "the Marshall merchants extended their trade territory."

The French prefigured all this. Having established and abandoned Fort Orleans within sight of Miami in the 1720s, they decided, twenty years later, to try again, and built Fort Cavagnolle. This fort they placed one hundred miles west and upstream near Kansas City.

My brother too has a story of the bridge. When he married, our father helped his getaway, driving him and his bride from the church in Marshall to a car stashed in Miami. A parade followed them from Marshall, George having many friends determined to mount a shivaree. The bride and groom got to their car at Miami and headed north. They paid their toll and crossed the river. Only one car of more than a dozen was that determined to stick with them. By that time, the toll was fifty cents.

Mother and Her Boats

There they all were, my father, uncle, and aunt, farming, digging in the dirt, and crafting archaeological reports, while my mother kept her eye on the river. One could not apportion the humors among them; three were of earth and one of water. Three were from Missouri farms, one from Chicago.

"I'm the only one without a book," my mother remarked on a recent visit, fixing me with her midnight eyes, totally unlike mine but the source of George's. My aunt, uncle, and father all have books, chapbooks, archaeological reports, and edited collections. Their subjects include the Spiro Mound, Native American pipes and bows, frontier trade guns, frontier photographs of the Sioux on the Rosebud, a year in Poland, Arrow Rock, and the Santa Fe Trail. Mother typed my father's work and joined the archaeological activity, not to mention the lengthy, unfolding conversation among the four of them. But she wrote no book. A Chicago girl, she stood apart from it all. She organized the first schooling in the county for children who required special education, beginning with a neighbor girl who had cerebral palsy, and she helped organize the first adult "sheltered workshop." But when you are among writers, it takes a peculiar kind of poise to endure having no book yourself.

She dreamed of a house on a hill, though it took years to come about. Bottom farms have no hills. Often, nearing Miami, we'd pass a knoll with a stately grove of trees, and Father would announce it as their intended site. Then they found the spot in Miami, a few acres of bluff and hillside above old ship landings. A principal feature of Mother's design was waist-high counters between the kitchen and the dining area, which flowed into the living room. She disliked being cut off from conversation while she cooked, and she wanted to step easily to the long picture window overlooking the river.

As a girl in Oak Park, Mother had taken piano lessons. Once a fire at home, a few days before a recital, gave her dress a suspicious smell. Shortly after Mother had taken the stage and begun playing Chopin, a breeze through an open window stirred the fumes from her dress and a teacher sounded the alarm. Mother fled the stage and never mentioned it until I pried years later, seeking stories.

She spoke softly and looked away from the dining table next to a baby grand piano given to her by her mother when she turned eighteen. Hardwood with teak veneer, it had been a significant present. By then, her mother had been a widow of several years and was teaching in the public schools. Mother stopped playing when I was very young. I don't remember her ever playing just to please herself. I suspect that when her mother-in-law, my grandmother the piano teacher, came to live with us, Mother became shy of the piano.

She might accompany Christmas carols, or lead "Happy Birthday," but I remember no more. Still, from the bench, she could watch the river. Or she could turn around and look at a wall of oil paintings by Grandmother, Tante Magda, and her father.

In 1951 levees were more rudimentary and the flood battle more in the river's favor. My father and uncle were leaders of the fight. Day and night they stayed in the bottoms, patrolling the levees on foot, seeking the weak points, finding a groundhog's hole, perhaps, and filling it, or identifying a spot where a crest threatened to top the levee and reinforcing it with sandbags. Women brought the food in relays—baloney, liverwurst, and ham sandwiches, coffee and more coffee.

Meanwhile, the road got worse. Water had crossed it in places, turning it to mud. This was long before every farmer had his pickup and before we had an army surplus jeep with four-wheel drive. That's when Mother drove up in our Plymouth with another load of food. Perhaps she didn't fully recognize the risk and so found it easier to take. "How did you get in?" Uncle Henry was astonished.

"I went kind of sideways," Mother admitted.

A few years ago, I was courting my wife-to-be, visiting her home in western North Dakota. Coming back on secondary roads, we drove through Sioux lands. There had been rain and construction, and at one point the road was out. We were not far from Wounded Knee and intent on reaching that site. We saw no sign warning we should consider a detour or turn back. As we came upon the break in the road, we could see a track off to the right through a field that in better weather workers and local drivers had taken. It was slick, or worse, and a worker warned that if we got stuck, no one was around to help us nor had they any spare equipment.

Dusk was coming on as I got out of my Toyota in the drizzle and walked along the track, trying to judge the ground. After a half a quarter or so, I paused and scanned the ruts that rose up over not an eminence but a true foothill, and fell from sight. Then I walked back, got in the car, thanked the worker, turned to my bride-to-be, and said, "Let's try it."

It didn't prove that hard. A kestrel hovering at the roadside before perching on a power line seemed a good sign. Kestrels had become our totem. We hovered for a moment ourselves, slithering uphill through soft spots while I gave it gas enough for momentum, and for traction, without letting the tires spin. For an instant I thought of Joe Chevalier sinking his tractor to its axle and remembered we had no neighbors here. But we made it, going "kind of sideways." That's a tender moment in my wife's memory, and I suspect it has something to do with my winning her hand, though I had proposed and she had said "yes" a week before. She credits my having a feel for the land. I suspect that I profited more from being my mother's son.

Like Mother, I grew up asthmatic. But she had few medicines available and sat propped up in bed some nights, wheezing, waiting for dawn. One relief was to leave the city and so, as a teenager, she found work as a nanny for a family with a vacation home on northern Lake Michigan, a position she kept for several summers. Once we needed to drive from Wisconsin to Michigan and took the ferry. That must have been unnecessary but a craving, more satisfying than chocolate. Briefly we were out of sight of both shores, which I could not get my fill of not seeing.

As a young man in Chicago, my father joined a boat club and took up canoeing. So he met Mother. They made a Canadian canoe trip their honeymoon. They drank lake water dripping off the blade of the paddle and fished for supper.

I will never know how much my mother was responsible for several of my own boyhood fixations—the navy over the army, Chris-Craft advertisements in the backs of *National Geographics*, the midshipmen at Annapolis throwing high their white caps. I remember no coaching and seem to have found them on my own. Clearly though my father knew what he was doing when he sought out the house site in Miami and cleared off the face of the bluff for a view.

The earliest vacation that I can remember was to Bayfield, Wisconsin, the departing point for a wreath of islands in Lake Superior. We still lived near Chicago and had a week or two on the lake with Mother before Father could join us. We kept a small cabin with a kitchenette and spent days on the shore in the sand. Mother had studied to be a primary school teacher, and she knew how to play with her boys. I was either in kindergarten or about to begin. The world around us was not rich in toys. We brought few with us, or perhaps those we brought were of little use on a beach. Lake Superior was cold, and we didn't swim yet anyway. So we built boats.

Mother found scrap pieces of two by fours, dowels, and laths. Where did the tools come from, the innkeeper? We sawed v-shaped prows and sterns onto the two by fours. We cut smaller pieces to fill out upper decks. We cut dowels for the stacks and nailed everything together. The bay was our ocean and then the beach beside it when we shivered from wading in icy water. The week seemed ocean-wide and we played as contentedly as passengers in posh quarters.

This is at the very threshold of memory for me, a spark separating from the utter dark, but Bayfield had a lighthouse. It jutted high from a bit of rock, just island enough to afford it footing at the outer limit of the bay. Bayfield was not a crowded resort area, at least not at that time. I remember no other guests, no one else except for the lighthouse keeper, who took an interest in a couple of boys eager to push homemade boats through the sand. I can't see this man at all, but he maintained the lighthouse and kept a boat by which he motored back and forth.

Did we clamor for a ride? Did we force Mother to make the suggestion? Was it her idea in the first place? Was the keeper more interested in her than in us? All I remember is boarding his boat—which may have been a Chris-Craft—and wearing a heavy sweater, and the boat's lift against the rushing waves as evening closed around us.

What would be the point of going to the lighthouse during daytime? The keeper himself wasn't there all day; his commitment was to the light he cast on the dark. I remember climbing steps from his wharf and being shown how those great beams streaked the sky and watching, mesmerized, as they played across the heavens. Later when I first saw the Northern Lights, I imagined a mythic lighthouse keeper.

When Mother and Father moved to their home on the bluff, Mother became the keeper, not of a light but of an old steamship bell that Father had found and mounted in the backyard. Every time she heard a tugboat laboring upstream, pushing its double train of barges, Mother would hurry outside, ring that bell, and wave. Suspended above her, the bell could encompass her head and shoulders. It seemed forlorn, an emblem of a world that she stood beside, binoculars in hand, watching it pass while straining to catch the name of the boat. She learned them all and soon discovered which crews were likely to wave back. Gradually she trained the pilots to expect her, and before long several gave warning toots so she would have time to step outside.

One day two men, having docked just upstream, walked up the hill to find out who had been waving to them. Mother served coffee and lemonade and learned about their boat. That's when she learned that they couldn't hear her bell above the racket of their engines.

Meanwhile she kept her log. Recently I found it in a kitchen drawer, 205 loose-leaf pages in a binder reinforced with duct tape, listing dates, the names of passing boats, notations of whether they went upriver or down, and remarks like "missed it" or "too dark." The log begins tentatively in 1962 on the backs of two scrap pages, but by the mid-sixties, Mother was logging up to fourteen pages a year, single-spaced entries in a hand that I now realize mine resembles. Then her entries dwindled to six or seven pages annually and then to only one each in 1982 and '83, to the last of which she added, "did not finish" two-thirds of the way down a single page. But she picked up again, and except for 1991, an aberrant year in which she recorded nothing at all, she filled four or more pages each year for another fourteen years. Mother kept her log for thirty-seven years, and that turned out to be her way of "building a book."

The barge season usually begins in March and runs through October. Once it began in mid-February and a few times in January. One year, she indicated December 19 as "End of season" then recorded the *Vicksburg* pushing six empty barges upriver on January 5 before setting up a new page for a new year. In the earlier years, occasional notes record identifying features—"two stacks," "orange paint"— whereas in the later ones she recorded function—construction barges, liquid carriers, and so on. *Vicksburg, Omaha, Lauren D, Dakota* (without the "h," a superstition of avoiding six-letter names having faded), *Brownville, Dixie Express, Jennie Dehmer, Sergeant Floyd, Tara Ann, Fort Pierre, Melinda B., Martha Trotter, Cecelia Ann, Brother Collins, Cindy Sue, Bull Frog,* and *Robin* are a few names, along with "up," "down," "after dark," "signaled," "tug alone," "first year," "after dark," "missed it," "missed it," "tooted," "waved," "flashed lights," "missed it."

Why did she do it? She scratched her head when I asked. "Oh, you'll never find an answer to that," Father finally said as I pushed the question and Mother sat there, wondering herself. "She's a deep one, your mother."

Her habits connected her to a century when steamboats had served as the pendulum of Miami's slow clock, their steady swings up- and downriver marking the essential movement the town surveyed. The old City Hotel advertised itself as standing "in full view of the river." Most of the travelers who put up in it had come off the river, and the whistle of the steamer rounding the bend warned them to finish packing, to hurry their breakfast, and to get back down to the docks to continue a risky passage.

In April 1852, the *Saluda*, loaded with Mormons, labored upstream against a rising current. Having reached Lexington, she could manage no farther, and for two days the boat, its captain, and crew waited impatiently, moored to the landing. That's when Capt. Francis Belt decided to risk it. He ordered steam and then more steam in his boilers, and when warned that the engines were putting out all they could carry, he ordered still more, declaring, "I'll round the bend or blow the boat to hell," so it is said. Captain Belt climbed atop and directed casting off, but within moments of testing the current, his boilers exploded. Belt, the first and second pilots, and twenty-four passengers died. The explosion injured many more. The boat's safe landed two hundred feet away on the bank of the river.

An earlier *Dakotah* blazoned that "h" because of the superstition, begun, perhaps, with the *Saluda*.

Between 1838 and 1887, a dozen boats foundered around the hundred-mile Big Bend, a "Steamboat Graveyard" that defines Saline County, among them the *Dart*, the *Euphrasie*, and the *Malta*, a "side-wheeler," that hit a snag in August 1841 and sank within a couple of minutes in only twelve feet of water. If loaded with furs, it must have been sweeping downstream; if wool coats, it labored against the current, supplying men on the frontier, and I'm guessing the latter, since that's when more steam would have been required. The *Tropic* sank a little upstream at the cost of nearly thirty lives. Mark Twain had once been its pilot. "A watchman's carelessness" caused the burning of an icebound *New Lucy*. The *J. H. Oglesby* sank at Euphrasie Bend, giving the *Euphrasie* something else in common with the *Malta*. Confederate troops burned the *West Wind* at Glasgow. The *New Sam Gaty* ran out of control in June 1868 and into the bank, struck a projecting log, "threw the boilers down," and caught fire. But during the 1903 flood, a Capt. Bill Heckmann took *Grapevine*, "loaded flat," that is with its gunnels at the waterline, and "saved 200 lives, two million of property and people from their housetops, working downstream from Rocheport to St. Louis."

Steamboats have given way to tugs guiding trains of barges up- and downstream, but the river has never lost traces of the exotic. In June of 1967, an engine room chief of the *Brownville* wrote my mother that he always turned a flashlight on her house when he passed in the dark because he knew she kept watch on barge traffic. Doug Frazee, whose father was born in Miami, remembers his tug's being the first to push six barges as far as Sioux City, Iowa, and of feeling that they "had accomplished something." A pilot no longer has to "read" the river as Mark Twain described learning to read the Mississippi. The channel won't have changed every time a boat jour- neys up or down it. But when the water is high and levees topped and lost as markers, some sense of the old unpredictability of rivers returns. It always commands my attention to watch a tug turn a line of barges, two and three abreast, completely around in the current.

If moving downstream, the pilot will aim the lead barges toward the bank, not too close, but out of the channel and into quieter water. Then the swifter current of the channel sweeps the tug itself down- stream, so the whole line pivots like a teeter-totter going up and over, at which point the engines bite into the current, the tug tilts its lead barges back into the channel and churns against the long resistance of the river. Then as it throws on all its power and edges into rushing water, I am reminded of the *Saluda*, whose captain called for more and more steam trying to get away from Lexington but blew himself out of this world instead.

According to the laconic newspaper account, "The surviving chil- dren of the passengers who were killed were adopted by Lexington citizens." No organized bureau of child welfare came to the rescue in 1852; there were no authorities to summon. The citizens did what they could, including initiating an unspecified number of young Mormons into the mysteries of becoming Methodist or Baptist. No doubt women like my mother had taken charge.

Chartreuse Sails

Charlie asked whether I got morning sickness. He was one of several black men with whom I worked on a fence construction crew one summer in Virginia. He was the oldest, the most experienced, and our foreman. On a clear day, the Blue Ridge defined our western horizon. "Of course not," I said, and must have looked startled for Charlie assured me that he "always did." He had just completed measurements and was pointing out where I should back up the small tractor, with the power auger on its rear end, and commence digging a hole. Charlie was as undefensive about his admission as the blue sky overhead. He stood smiling, open, and armed with a manual posthole digger with which he would clean up the rough plunge of the auger about as quickly as I could lift it out of the ground and maneuver the tractor into position for the next hole. I spent a fair portion of that summer trying to mimic the perfect efficiency with which Charlie used his digger, the sure vertical plunge making a smooth cut shaving down one side of the hole, the quick, horizontal tug on its arms, closing the jaws below, then the clean jerk of clay and loam up out of the opening, leaving the hole smooth and ready. Place two holes together, line them with a blanket, and you could bed down a child.

That night I dreamed of lying not in a posthole but in a furrow, just beyond the shade of a pair of basswoods I had passed repeatedly, and of giving birth. I lay on my side and bare bottomed, loam cushioning my cheek, while I sought protection in fresh black earth unlike any in Virginia. But a furrow's not that deep, and I lay exposed, the sun warming my backside while new wheat gleamed at eye level. Morning sun streamed across it, washing it chartreuse while I pushed a baby out onto the land. "Sails," I thought, "chartreuse sails," as tiny leaves of new wheat billowed on my low horizon.

I was young, and my young wife was pregnant for the first time. But the ground in which I was lying was not in Virginia, and the tractor that had prepared it wasn't the little Ford with the posthole auger.

The tractor was a John Deere G—a Popping Johnny—that I had once worked back and forth across a hundred-acre field, a tree line defining the east end and a dirt road the west. Three fields lay to the north, two more and our strip of timber to the south. I plowed the

Middle Field on a March day of the sort that brings out T-shirts and Frisbees on a college campus. Again I stood at the wheel. The east end of the field gleamed in winter wheat, new, tender, and brilliant as a June lawn. I plowed corn stubble that abutted it. My tractor pulled a simple two-bottom plow. Two coulters cut into the ground, then two shares followed, knifing deeply, peeling back, and rolling the earth over, earth as black as a moonless night. The tractor moved slowly, and I spent most of the day turned half around, watching the staggered peelings of earth roll over. I saw the plowshares as hands sliding under the earth and lifting the dirt as it turned, broke up, and resettled so that another furrow lay stretched out naked and revealed.

I was mesmerized by the widening strip of black earth against the lime-lime green of new wheat in spring. We could have named our daughter "Chartreuse," but we didn't think of that. I think instead of how visible work, with land, machines and tools, has nudged me toward mysteries, often feminine, and of how the boyishness of trying to master such work knocks at their doors.

When we cultivated, small green corn glimmered down rows where in other years wheat had grown and in others water had crested for miles in all directions, higher than the stacks on my tractor. Then clumps of trees broke through like whispers of islands. Cultivating, I worked four rows at a time, with "shoes," small spear-shaped blades, breaking up the dirt between each pair of rows. My small front wheels jerked and pitched over the clods that remained after plowing, disking, planting, and the earth settling to a seedbed still rough enough to suggest waves lapping against a bow. If the tractor followed a sudden twist of my wheels, if I didn't correct quickly for its jerks and pitches, I could suddenly take out rather than clean up four rows of corn.

The tractor moved at the pace of a brisk walk, east four rows then west four, and at the end of each round I could skip four rows to make the turn easier. Turning required tripping a lever behind me to pull out the shoes, braking the wheel on which the turn was made by pressing down on a pedal, spinning the steering wheel and lining the front wheels up between the second and third of a new set of rows, and dropping the shoes back in. Skipping four rows and moving to the next set down the field made those several maneuvers easier. I could work back across the field later and pick up all the skipped rows with the same smooth motions.

As I became more agile with the tractor, I found it beneath me not to make the tightest turn possible. I'd cut the throttle with my right hand, pulling the cultivator lever behind me simultaneously with my left, then disengage the clutch by jerking back a second lever on my right, brake the large left wheel with my left foot pressing hard on the left foot brake, spin the steering wheel left with my left hand while riding the clutch with my right, line up the very next four rows by swinging the front wheels smoothly around and between the second and third of them while the huge left wheel stood almost still, turning on some small, unnoticed chartreuse sail, reset the cultivator by pulling on the lever behind me again, watch the shoes slide into the earth at the very starts of new rows, throw the throttle forward triumphantly and fully engage the clutch. Then with a smile that I'm glad no one was around to record, I could loaf and invite my soul.

Years later, George, the father of three daughters, said he had just about the same dream every time.

River Horse

"A tree's just a weed if it's in the wrong place." By now you recognize the voice, and if you grant the fact of farming, he's got a point. Five-ten would be tall enough and no slouch in his bearing. A widow's peak and thinning hair, mostly dark until late in life, parted in the middle and swept straight back. Heavy cheeks, a mustache, the exposed skin on his face and arms leathery and well tanned. At work, he'd have a tool in his hand—a hoe, an ax, or a wrench. Usually a touch of irony twinkled in his blue eyes.

He made a play of strength, and I recall a photograph of him pitching a log onto a pile for burning. The log was six feet long and thicker than his thigh—strong man's stuff for sure, which he made seem light. But it was cottonwood, not balsa but tending in that direction, so the photograph was a joke. At the same time, Father would complain that Henry didn't know when to stop tightening a nut. More than once I had to dig around to find a foot or more of pipe to fit over a twelve-inch crescent wrench and gain leverage enough to loosen a nut that Unc had worked on first.

During clearing, a bulldozer pushed large piles of trees and brush together for burning, the cores smoldering for days. Piles would rise larger than houses, and Unc would stride up, splash gasoline around one, and toss a lighted march. Whoosh! I never tired of that moment. He'd strike the wooden match with his thumbnail and stand a few feet away from the pile, warning off the rest of us.

His care for small things was legendary, when they weren't weeds. Just as he picked up Spot during a flood and gained a dog, another time he brought home Timothy Tiger, a yellow tomcat. I've seen him swerve his pickup to avoid a butterfly. Then there was the time he found a groundhog's hole on the remnant of levee that edged our strip of timber. No longer a main levee, it could still prove useful, limiting the damage of rising water. No reason to let it fall apart. The groundhogs would have to go. Fill the hole in and they'd just dig it out again. They'd have to go permanently. We owed that not just to ourselves but to our neighbors.

Unc walked over to the machine shed and filled a gallon jug with gas. He walked back and knelt beside the hole. Carefully and lightly, with the side of his hand, he brushed away a line of marching ants and waited for them to move into the brush. Then he sloshed gas down into the hole, flicked a match with his nail, and tossed it.

"I hope the folks are home," Unc said, rising to his feet. You have to get the beat and his lingering over the Os: "i HOPE the FOLKS are HOME." We turned away and went on to other work.

Late one night, during a spring flood in the early '60s, someone found such a hole on the outer side of the levee, below the waterline and not easy to get at. Several hands were needed, some to fill sandbags, others to carry them to the point of defense, still others to get into the water and place them. Father and Uncle Henry, the two oldest workers, got in the April, nighttime river. Father took off his clothes, figuring he'd have dry clothes to warm up with when he got out. Henry just waded in. A local radio station had come for on-the-spot coverage and a reporter stood on the levee, bending toward Henry, reaching with a microphone toward his mouth while he took a bag, went completely under water to place it, then stood back up blowing river off his mustache.

I don't know what he said since I was off at college and so not there to hear. The line that lingers in general memory is Duke Cheva-

lier's, who, standing to the side, fifteen years younger and dry, murmured, "He's a tough old sonofabitch," which was his way of saying
"he's competent." I remember instead another line of Unc's, for no
special occasion, so it may serve for this. "The hippopotamus is the
largest, even-toed, non-ruminant, ungulate in the world, and what
does it get him." River horse ("hippo-potamus") that he was at the
moment, he knew what he was saying as he added, "We can rent
this river, but we'll never own it."

Father

The older brother, let us conjecture, could afford the faith of a classicist, because he was confident of adult attention. The younger, then,
might lean toward the romantics in his quest for a share of the same.
My father told me the story of Uncle Henry as a river horse. He put
it in a letter. I was half a continent away, and he wanted me to get the
picture. Henry was seven years older, and Father never got over a
degree of hero worship. He was in sympathy with his subject and so
heeded intuitions a classicist might suppress. One morning in 1984,
Father turned his car around and backtracked a couple of miles, explaining to Mother that he "just wanted to see Henry." He just "had
a feeling," which was rather unlike the scientist in him, the practical farmer, agnostic, and trained engineer. So Mother and Father
dropped in on my aunt and uncle and the four of them sat visiting
for half an hour or so; then my parents continued home. That night,
Henry suffered a stroke. He died within a month, and there was no
more conversation.

Conversation is one thing Father cherished. Uncle Henry often
spoke as if he were on stage or at least at the head of the table. He
must have cultivated being quotable. Father preferred to be engaged
and fully present, man to man, or man to woman for that matter, one
on one and even. For many years he belonged to a discussion group
of men, mostly businessmen and a few professors from our small,
local college who met to argue current events. He always had my
attention, and whenever I needed, I had his.

Or whenever he thought I needed. He enjoyed a radical period
as a young man in Chicago during the depression but became more
conservative as his responsibilities multiplied. When I was in college

and making the opposite turn, he never belittled me for it, but he did pay attention. We exchanged letters, once every week or so for quite a long time. I can't be sure of how long, but it seems like well over a year, and I'm the one who flagged first. They were four- and five-page, typed, single-spaced, argumentative letters in which he took heed of all I was trying to teach him and argued back, never letting it go unchallenged. My liberal professors might reconstruct me, but not without a fight. We don't have those letters now, and I'm glad enough of that; but I've always thought it more remarkable than a string of perfect growing seasons how he took the time—when I had much more time to burn than he—and listened, and argued head to head with me. An impartial observer would surely have given him the edge.

He is no taller than Henry and stocky. He radiates firmness of purpose. He was a "fighting guard" on his basketball team in the days when each score was followed by a center jump and high team scores seldom topped twenty. He has strong, broad features, high cheekbones, blue eyes, and a mustache. He always looks ready for a brisk walk and carries a full head of hair ready to blow in the wind. First it had a reddish tinge, then brown, then gray; now it is fully white. He has always called his father "Father."

He enjoys a martini, a bourbon, or a scotch. He favors scotch but isn't particular. I never remember beer in the house. And he has his drink, or two, before dinner, then quits. For dinner, or for any meal, he enjoys a glass of milk. He has drunk so much milk that, although he has fallen several times since turning ninety, he has yet to break a bone. The milk is always good, as is toast with jelly, a strip or three of bacon, chocolate cake, and shrimp. "Aren't these good shrimp?" he'll ask at a restaurant, any restaurant, without again being overly particular.

He loved his dogs, all strays, Spots, Nippy, Wags, Mut-tilda, and Bonny Dixie Ever Loving Jellybean. "How come puppies are so damned cute?" he asks, while balancing a treat on Mut-tilda's trained-to-be-steady nose. "Evolutionarily, I mean. What's the reason?" Not that he lacks an answer.

"You know the best thing about your Aunt Jean?" he asks, rhetorically. "She was a great lady, but her very best quality was that she didn't brag about drinking her coffee black." That came from his discovery late in life that he actually liked coffee when he added milk. Apparently his mother, father, and brother all drank their coffee black and frowned on those who didn't. So Father adopted the family ways until, at eighy-five or so, he tried cream and found he liked it. But then he always had a capacity to live by an experimental method and change his mind when offered proof.

He has always been fond of, nay courtly to women, and he always wanted a daughter. So he took special delight in mine, in George's three, and in our wives. He's a bit of a flirt too, and on my college graduation weekend he seemed effortlessly and continuously the center of attention among all the parents who gathered around my closest friends. His robust good health didn't handicap him at all, neither did his open curiosity, his readiness to tease and to be teased, his information continuously stored on numerous subjects, nor his experience first in but then much more profoundly away from their urban and suburban landscapes. In that New England setting, he was the frontiersman who had come to the Ritz and who instantly reshaped the ballroom simply by standing in it, wholly comfortable, and ready to talk to or, surprise, dance with anyone.

The most romantic act of my Father's life, courtship aside, about which I know next to nothing, was to join his brother farming. No doubt, more went into the decision than I will ever know. The story we grew up on was of Father's frustration in business, which led to ulcers for which farming offered relief. He was in purchasing, the service side of business. Leadership came from sales, which brought the money in. Father saw himself as a potential leader trapped in a storeroom. "If not a business success in ten years, slow down and enjoy life." That had been the first of my parents' goals written down during their second anniversary dinner, when Mother was quite pregnant with me.

So far so good. But farming didn't exactly entail deceleration. At forty-two, Father took arms against a sea of difficulties and waded into them. There was land to clear, a farm to construct, floods to fight, investments to make good on, flood control to fight for, archaeological salvage work to assist as at least partial acknowledgment that the control they sought exacted its price. Then there was the annual struggle of seeing crops to harvest and of uncertain income always, since there was no regular paycheck, just crops to sell when one had them. Given floods some years and droughts others, Father's move seemed better calculated to produce ulcers than to heal them.

But there was romance in joining with an older brother and sharing his goals, such as regaining a farmer's place in the world no matter the difficulties. They went back to the land long before my generation toyed with that notion. They diffused part of the strain by cultivating projects beyond profit. On many winter nights, Father developed photographs in his basement darkroom. I showed little interest in photography myself, but I dipped and moved the blossoming prints in their various solutions and set and watched the timer. He rarely missed a chance though to engage me in something. Once we skinned a raccoon we found dead by the side of the road and started to make me a coonskin cap. Too bad the tail broke in half, though that moment added a pleasure, since that's when I first learned that Father knew more cuss words than I. We shaved and shaved down another cured stave of Osage Orange, going to the basement again after many dinners, hoping to make me a longbow. We went too far in our shaping though and cracked it too. We made and flew a box kite and built a reflecting telescope.

The box kite reverberated with Father's own youth. He told of a time when kites suddenly seemed to appear in all the stories he read, but no one in his town had seen one. They weren't for sale in a drugstore. So one Sunday he tried to make one for himself out of sticks, string, and paper. He had made, or had tried to make, all sorts of things—a cart for his dog to pull, a crossbow, a harness—but his father and older brother were too preoccupied with "men's work" to take much interest. This time it was different. His father took up the making with him. He offered suggestions and help, and it thrilled my father for his father to center his attention on that project. It was like being ready for a night of meteor showers and finding that night cloudless.

So they made a kite, copying a picture in a book, and flew it. But the kite just kept diving to the ground. At length, my grandfather returned to his chores, and Father stored his kite in his room as a memento of one close afternoon with his father. Several years later, he found an article that refined his understanding of a kite's tail. So he got out his kite, tore up some rags, added sections to the tail, and "that evening, after supper," he told me, "the two of us took it out again and flew it until we could no longer see."

Our telescope began with grinding, shaping, and polishing a pair of concave mirrors. Mother and Father had begun the project before moving to Missouri. Then it sat in the basement unfinished for several years. One winter, Father got it out, and George and I continued under his guidance. We had an abrasive compound and a stone that shaped the concavity. We took turns at the workbench grinding, very finely grinding those mirrors. Eventually we positioned the mirrors on a beam so that the larger one could capture the rays from the moon, or Pleiades, or Saturn, and throw an image onto a second, smaller mirror positioned to direct it into an eyepiece. Then we took our telescope out one clear night to see the craters on the moon and the rings of Saturn.

At least half a dozen neighborhood kids joined us to see those rings. We saw two clearly. At moments, we felt sure we saw a third. When Father told us they approached 200,000 miles in diameter and were thin enough to be transparent, we ran off into the night astonished, our minds lit up like sparklers. Father tried to refine our understanding. Saturn was almost 900 million miles away. That's 600 million lengths of the farm. How many times might we walk that mile and a half in one day? Ten? Well, if we said so. Then we'd only need 60 million days to walk to Saturn.

Or what about those rings? The widest diameter is 170,000 miles, only 113,333.33333 lengths of our farm. We were down to 11,333.33333 days walking. He had us do the long division and pushed us to push the decimal. Its endlessness helped us intuit something about distance, the very large and infinitely small mirroring each other.

For distances thrilled Father. He felt awe and reverence, he said, handling artifacts hundreds, sometimes thousands of years old. All his scholarship began with hobbies. He had been an archer, and that led to *Native American Bows*. But he admired the longbow, and the recurved, sinew-backed bows of older times, and the craft of makers who lacked modern tools. When archery added sights, pulleys, and mechanisms for checking an archer's exact draw, he took up muzzle-loading rifles, saying, "You can't ruin those."

Collecting a few old guns led to making them work, and that led, in turn, to examining their parts. Before long, caches of frontier trade guns were being sent him from far and wide for analysis. His study brought him closer to LaSalle and Bourgmont, and he gathered and read an extensive library on exploration and the fur trade, a library that I have drawn on here, and he was partner to research projects in Michigan, Alabama, South Carolina that I know of.

The Black Sand site got him back into the field at eighty-eight, fingering a flint blade and pottery with the earliest decorations yet known in the valley. That site hinted at future surprises: the '93 flood had signaled it, what will turn up next? Only his persistence added the mastodon to the scientific record. Without him, those bones would have sat in paper bags on storage shelves for decades, perhaps forever.

Perhaps something is in error. Perhaps Oneota youths of comparatively recent time came upon a surprise themselves, mastodon bones washed out by a flood. Perhaps they hacked apart some mysterious ribs and tusks and carried their find to higher ground where time reburied it. Then again, perhaps not. Not only the bone, stone, tusk, and tools, but also traces of the loess they were embedded in yield very early dates. Challenging ideas awaken romance, and the notion that humans entered the Americas almost three times earlier than we had thought, that we don't know who they were, where they came from, or even whether they are continuous with later Native Americans is sublime rather than fanciful to my father.

In an unpublished essay, "The Testament of an Ex-Southerner," he taught himself to challenge equally the received opinions of Little Dixie. He discredited secession and the term "War Between the States," reasoning that the Union had to preserve itself. Later German and Soviet threats argued for our united strength. In the same essay, he acknowledged the slave-owning background of his Kentucky grandmother's family and found "obviously ridiculous" the rationalization that he heard more than enough, that black brains "ossified" at puberty. In his essay he guesses that his mother gave in to too many Southern atrocity stories.

He denies too the convenient fiction that Southerners treated their slaves like family. He cites an 1846 letter from a Virginia lawyer to a Miami resident. The lawyer advises "cashing in" on six slaves "at the appraised price" or selling them all to a trader. The lawyer is indifferent to the consequences of trade to the slaves themselves. He names one, "Fannie (a young girl about 11 years old)," whose feet were "very much frost bitten." The lawyer recommends curing the girl "as soon as possible," but only to facilitate her sale. Father read the lawyer's indifference, except as one might care for cattle, as "more than incidental cruelty," and he stood firmly for integration when it finally came to our schools.

Atrocity stories come from all sides, however, and some do tell for the South. Here's one such that Father collected. During the War, the Missouri Militia (Northern) caught a half-witted Miami boy and began to question him about (Southern) guerrillas. The boy would or could not tell them. It was winter, the river frozen over. Convinced that he knew, the militia took the boy out on the ice, cut a hole, and threatened to put him under the ice if he did not lead them to the guerrillas. The boy still would or could not tell, so the men pushed him into the water. When the boy caught hold of the ice and tried to pull himself out, one of the militia cut off his hands with a saber. After that they had no trouble pushing the boy under.

"The saber could be an artistic touch told to heighten the story," Father notes, "but the fact remains, they took the boy out on the ice and he never came back."

It's too bad Father couldn't have come upon the militia that day, for that's one atrocity he might have stopped, at least according to my uncle. For he told me once of a time the two of them were driving near their homeplace; Father was a teenager, and the car, driven by Henry, was their family's first. They came upon a crowd of men and boys at a bridge. Father jumped out to see what was going on and found a boy, two or three years younger than himself, poised on the girders, frightened but about to jump, the crowd egging him on. This angered Father who called down to the boy, "Don't be a damned fool. Climb back up here," and much relieved, the boy did. Then as the boy was extricating himself from his perch, Father turned on the crowd and said, "If you all are so brave, let's see one of you jump." Like cockroaches under sudden light, they dispersed. At least that's what Uncle Henry remembers, and that it was pretty much all over before he got out of the car and over to the rail himself.

In most cases Father thought for himself, and he tended to be a dissenter. Within a few years of our moving to Marshall, he was elected to the city council. An issue at the time was "widening the square," as was said, though that really meant widening the streets around the square and so cutting down a number of specimen trees, oaks and maples. Forget that we were clearing land to farm at the time. A farm is one thing, a county seat's square quite another. Father fought hard against the proposal but lost. Years passed, a decade or more. By then he had built his house on the Miami bluff and had almost immediately become Miami's mayor. (I lived in Barranquilla, Colombia, at the time and had great fun telling my students that my father was Mayor of Miami, and, by the way, that my brother was George Hamilton, who kindly wrote letters to verify the fact.) Then the Marshall Council proposed to widen those streets again, to provide still more parking spaces around the square. But as Father was quick to realize, that meant taking land from the county, not just from the square, for the square with its courthouse was county property, and offering nothing in compensation for it. The council, he claimed, was proposing to act as if the square were the private property of the merchants around it. He wrote a letter to that effect, circulating it to all the other smaller town councils, and that ended that.

A year or so later, Mother and Father found they needed a new refrigerator and went to a square merchant to buy one. They hadn't been in the store long, looking possibilities over, before the owner came up and said, "Ted," for that is his name, "when are you going to die?" Probably he meant to joke, but no one in our family is quite sure. In any case, Father has outlived him.

Farm

To farm is to hold onto something, and a farm is land to grasp. Its Indo-European root has to do with affirming and confirming, positive assertions beneath which one senses the infirmity of a grasp that weakens. "Farm" comes down from those old roots through the Medieval Latin *firma*, a fixed payment for rented land, a farm. In Spanish, *la firma* becomes a signature, a guarantee, a signing on.

Standing on his tractor, Uncle Henry was grasping more than the wheel. He had grown up farming. He had studied agriculture at the university. He had spent a critical year in Poland and Russia after World War I, helping reestablish farming in a broken land. He had come to Saline County as the representative of the University of Missouri Extension Service. He had worked for a bank as an assessor of farmland. But he was not then an owner of land, not an affirming farmer. His parents, meanwhile, lost their farm. Their *firma* had lost most of its value. Then my uncle and father started over, beginning where they could.

Again and again my father spoke of how much he valued "being his own boss," able to make his own decisions and work through his mistakes. The relief he felt for what he had gained seeped from him like new meanings from old words. But I never heard Uncle Henry express such feelings. Instead I see him standing upright, his hand firmly on the wheel, the tractor performing the primal act of breaking into the land, readying it for seed, and so acting out his affirmation over and over, as if to reaffirm his parents.

My father plowed too and described it lovingly.

> In a good sandy loam it is just plain fun, especially when you are plowing under wheat stubble and the sunflowers have had time to grow. I never got over my astonishment for there they stand in front of your tractor, eight, ten, even twelve feet tall and so dense you can't see through them. But their turn comes and the plow flops them under, and you look back and not a trace of them is to be seen, aside from an occasional leaf sticking out from under a clod.

"But their turn comes." I note the satisfaction in that, and the magic, too, not the least of which was seeing that sunflower flop under the soil like any superior who had got, as Father would say, "too big for his britches."

The two of them had grown up using mules like Frank to pull single-bottom plows—that is, plows with a single cutting share. They had crossed into an age of tractors with double-bottomed plows and then larger disk-plows that moved faster, tumbling milkweed, sunflowers, and pigweed—"careless weed," we called it, and whether it was careless or could care less, I never knew—turning all that under swiftly while turning up dark, workable soil. Each spring they affirmed another season and another chance as the hard-worked, forgiving earth welcomed seed once more. Plowing cleared the ground for them in more ways than one. Alongside the old prairie, it served as prayer.

The loneliness of the work adds to its meditative aspect. A medieval farmer would have rarely worked alone. The stark aloneness of farmwork is recent. Through the turn of the twentieth century, Hamlin Garland wrote midwestern stories in which roads were full of people walking to and from their fields. In a Garland story, a farmer steps from his door and within moments meets neighbors. People stride with him or pass by in wagons. He'll hoist himself onto the back of one and ride along with his legs dangling over. In my youth, the likelihood of walking from the Old Fort to the farm without a neighbor coming by and offering a ride was much less than even. Old section maps indicate school sites within a mile both north and south of our farm. Schools meant families, people in the plural working the land. Most of them did not hold onto it, however. Few had much of a grasp.

Those one-room country schools have been closed for more than half a century. My father and uncle employed only one farmhand in the early days, to whom they also offered a small house. Later they only brought in temporary help, often enough us boys, until George became the third farmer. Later he and a brother-in-law worked alone—just two men and large, expensive, efficient equipment farming over one thousand acres. Any one of them standing at the wheel of a tractor was a symbol of the lay of the land before us.

I won't echo Melville and say we kept "sorry guard," although it was imperfect. In 1900 the population of Saline County was 33,000 people, nearly all of whom farmed. Now the population has dropped to 24,000, and half of it resides in one town. The most ecologically enlightened approaches to farming would require a fair number of people living in cities and towns to return to the fields, to reduce our reliance on chemicals and our consumption of fossil fuels. I hear no clamor to return.

Nor have I volunteered. I doubt that I would have written this book had I fully grasped the farm. Recording what I have of it, and of our neighborhood and its history, has depended on my getting far enough away to look back on it, much as Father spoke of plowing not from the tractor seat any longer but from remembering having plowed. The jolting eases, as does the dust and tangle of weeds, the share broken on an old stump and a day lost to unmounting, welding, and remounting it. So does the anxiety of planting quickly while the weather holds and of supporting the unwieldy investment in time, seed, machinery, and materials that will show no return for months, if at all, much less the work in cold, gloomy sheds through the winter, coaxing old equipment through another year. "What would you do if you won the lottery?" runs the joke of farm country. "Keep on farming, till I lost it all."

My brother no longer farms and does not miss it. Quite a few of my high school friends have left or have been forced out of farming, and I can't find one of them who laments it much. The usual response is, "Why didn't I think of this sooner?" "Why was I ever fool enough to farm?" or "Sure I'll farm again, when I retire." And though I haven't heard that last remark from George himself, I notice that he now finds numerous reasons, far more than are merely practical, to drive alone into the bottoms, by the old farm, and through the wildlife area, thereby taking a roundabout way to or from home.

Beyond all the economic problems, today's farmer bears the brunt of criticism for a century of questionable land use, born of another time and discipline. He stands a little like the Vietnam veteran who carries home the stigma of having participated in a questionable war. There is no question that the land has a different character from when we began. Only scraps of timber remain, and most wetland is gone. Among farmers in the bottoms, as in wetlands elsewhere, farming means drainage, cutting ditches and setting culverts so the fields will dry out enough to work. Among my university friends and others, that has become a transgression of the worst order; those being the same people, often, who find transgression thrilling when it is their own adventure in science or in art.

An enormous volume of chemicals has gone into the soil. Agricultural runoff damages the groundwater and reduces the number of worms. Even with the minimal erosion in the level and leveed bottoms, much has blown onto the river, down to the Gulf, not to mention what comes and goes with the floods. The river has been rip-rapped, "stabilized," channeled. It can seem a canal. Huge dams hold water back in flood season, but when a flood gets out of hand, the levees, ever higher, always stronger, lift the crests and confine water that otherwise would have spread out earlier and with less force across wide bottoms. Levees channel and aim the river at more destructive chances.

At the same time, the Tetesaw Levee, like many others up and down the rivers, held all through the sixties, seventies, and eighties, allowing the development of more efficient farms. Fewer workers labor in the fields; many more students complete high school and college. My early school years were scarred by numerous classmates dropping out after the eighth grade, which was as soon as they legally could.

Most dropped out to farm, and quite a few other emerging farmers failed to complete high school. Neither my grandfather nor great-grandfather Hamilton were high school graduates, and few adults suggested it should be otherwise. Whereas George and I assumed we would go to college, it had been doubtful for our father who dropped out of school twice. Once during high school, when heart disease crippled his father, our father, at fifteen, took over their farm. Then about five years later, halfway through college this time, he dropped out again after a tornado leveled his parents' home. Only the insistence of his mother, a Chicago woman, got him back to college to graduate in time for the depression, and to the luck of a job in Chicago, where he met my mother and prepared, at great length and indirectly, to return to farming almost two decades later.

Mid-twentieth-century photographs of eighth-grade graduation in country schools suggest a serious and dignified commencement. Young boys, farmhands already, stand soberly in their coats and ties, well aware of their coming duty in the fields. Such a graduation could easily have been the only one for my father and uncle. Then through that period and beyond it, as farming modernized and those same students went on to high school and college, the cost of food as a percentage of family income, and as the percentage of the nation's work time given to obtaining it, became lower than for any culture ever, except the Lotus Eaters.

To rush along dragging much with us is human. So it seems. That was the vision of the last century, which begs for correction. Earlier the vision had been equally aggressive—small cities strung all

along the river. The means though for enacting it were less available. Indians had had their turn, layers and layers of them, like the
fertile soils laid down by the floods. And so a last story of the Missouris, from their time on the banks, and caught in the current of one
deep river.

Deep River, a Conclusion

Jean-Bernard Bossu's *Travels in the Interior of North America, 1751–1762* records the story of an early French trader who visited the Missouris. Even earlier an unnamed hunter had introduced the Missouris to firearms and had supplied them, to his benefit. Now this second trader wished to sell ammunition, but their need was not yet great enough. So he told them that the French, who called powder *grain*, sowed theirs in fields and harvested it like millet. Accordingly, the Missouris sowed all that remained of theirs, posted a guard to keep off marauding animals, and resupplied themselves from the trader, "who made a considerable profit in beaver, otter, and other furs" and paddled away smiling.

Not long after, a colleague returned to do more business with the obliging Missouris, who welcomed him warmly and even granted him the use of their common lodge to display his wares. When he had laid out everything, the Indians started milling around in confusing and distracting ways until they carried off every single item he had brought. So the trader appealed to the chief and complained bitterly of the unfairness of the Missouris. The chief replied with great dignity, much as a river tops a levee, that his people would repay the trader well, they would swamp his boat with furs—as soon as they had harvested the powder that they had sown according to French instruction.

Bossu is the writer who mentions the Missouri Princess living among the Illinois after becoming the widow of Sergeant Dubois. It was to him that she showed the "repeater watch set with diamonds." Like many a traveler, and raconteur, Bossu could mistake the curious for the important. He notes the watch and that the "chief, or Indian king," of the Illinois at the time of his visit "is the son of the one who went to France with his entire retinue in 1720." That would be Petit Missouri, Bourgmont's son, who should then have been around forty. But Bossu does not recognize the woman as Petit's mother.

Perhaps sojourning among people with whom he failed to identify, Bossu could not recognize affections we cherish and take to be ours alone. Perhaps long absent from home himself, he had suppressed them. Or perhaps the woman was not Petit's mother. In Bourgmont's dozen or more years among the Missouris, he may have taken one woman then another. Perhaps the woman I have been calling the Missouri Princess was a younger sister, Petit's aunt. Perhaps she was another tribal member.

Or perhaps not. That she should be Petit's mother is made more probable by Petit's having found her. We can imagine that news of her came to Petit much as that Ithacan farm boy stumbled on Odysseus' oar, and he made his way downstream with it until he found her and a chiefdom. So he engaged the Illinois, all for love of his mother, a Missouri woman who had traveled as far as Detroit and Paris and who had passed by Chicago.

It was not much of a chiefdom anymore. The Illinois, who had given the Missouris their name, were no longer powerful in the river valleys. But Petit moved competently among them. Perhaps his tie to Bourgmont and whatever inheritance he had been given—a fusil, physical stature, tales of Paris (quantities of meat on the Rue des Boucheries, sorcerers and magicians at the opera, the "lodges" of Versailles, effeminate, "alligator-smelling" gentlemen)—perhaps all this served Petit Missouri well among the Illinois. Perhaps his experience with the French made a chief of him. The Illinois still had their place, at least a minor eminence, and Petit Missouri's prowess had pulled him up on it.

Then since the Illinois were closely related to the Miamis, as the Missouris were to the Osage, the Miamis' new connection to Petit Missouri may have led them into old Missouri country where they would name a town a half century or so later. It is a town to which I helped move my parents and one just outside of which my brother built his own house on the bluff. I have never lived there, but for forty years I have come back and back to it, always calling it home.

Acknowledgments

This is not a big book, but it covers a lot of ground, on little of which am I expert. It's true I wrote it, but that odd fact calls nicely into question just what "writing it" means. So much of it I heard before; some I witnessed. At times I felt of the various voices in my ear that I was but their amanuensis. Which is no false modesty, I did my part, but that surely is part of a much larger whole.

For obviously this book would not have been possible without my drawing at will on my family's long commitment to farming, as on their involvement with Missouri archaeology and with the history of Miami and the Tetesaw neighborhood. For that and much more, I am ever in their debt. Their work led me to the archives of the Western Historical Manuscript Collection, where I discovered one vignette after another. James Thorp's eight-volume scrapbook is there, and for all my years at a university, I was never more the scholar than when plundering it and boxes of letters, photographs, and old newspapers in those exceedingly pleasant, pleasantly staffed rooms in the northwest ground-floor corner of Ellis Library in Columbia, Missouri. A smaller collection in the Marshall Public Library aided my efforts, along with tapes my father and uncle made interviewing Sam Irvine and Paul Stonner. Then, too, there were the books in my father's library, several of which have migrated to my shelves.

Time and time again, my brother, George, and I have driven through the levee district speaking of one part of it or another. In many ways, this book is as much his as mine though there may well be parts of it that he would not claim. My cousins Jim and the late Anne Hamilton, all but unmentioned here, shared this life, too, although less centrally from my perspective. My agent, Lynn Franklin, has been a most influential reader. Her fondness for this project from the start, coupled with her resistance to earlier drafts of it, led to

better writing, for which the reader should be grateful. John Rae-
burn, my longtime friend and from-the-beginning colleague, read
the entire manuscript and made many comments, all of which im-
proved it. My sister-in-law, Lee Hamilton, my former wife, Antonia
Hamilton, and Jenny and Colin, our daughter and son, read, heard
told, and occasionally lived along with moments depicted here; and
Colin has read the whole manuscript, some of it more than once.
A writer himself, his reactions have been sage and accurate. Keith
Achepohl, Stavros Deligiorgis, James Feathers, Laurence Goldstein,
DeWitt Henry, Mary Hussmann, Howard Junker, Carl Klaus, Joseph
Mullin, Laura Rigal, and Steve Weiland also read, listened, ques-
tioned, encouraged, and inspired.

Mary Jo Berry and Jamie Nichols, Saline County's Recorder and
Assistant Recorder of Deeds, and the staff of the Saline County Title
Company helped me get my bearings. The University of Iowa provided a research grant and a semester's appointment to the Ober-
mann Center of Advanced Studies, which permitted much of a first
draft. It is hard to remember now how much of this was rehearsed in
the coffee room there, but some surely was, and I learned from that
give and take, as I did from seminars sponsored by POROI (Iowa's
Project on the Rhetoric of Inquiry). Nor should I forget the editors of
the *American Voice, Country Journal, Creative Nonfiction, North Dakota
Quarterly,* and *Poetry Northwest,* who have hosted, at one time or another, bits and pieces of this work, usually in a different form. A few
pages of it are also taken from *A Place of Sense: Essays in Search of the
Midwest,* edited by Michael Martone and published by the University of Iowa Press, 1988. I am also much indebted to Beverly Jarrett
and the staff of the University of Missouri Press, especially Maurice
Manring and Sara Davis, who guided me through the crucial steps
of acquisition and book production.

Finally there is my wife, Rebecca Clouse, my companion in reading, writing, and more. It takes years to "build a book," even a small
one of no great distinction. Her imaginative intelligence, patience,
sharp eye, and attentive ear, not to mention her loving friendship,
are not at all bound by these pages though they enhance each one.

Bibliographical Appendix

Throughout

History of Saline County, Missouri. St. Louis: Missouri Historical Company, 1881.

Miami Index. State Historical Society of Missouri, Columbia.

Miami News. State Historical Society of Missouri, Columbia.

Qui Vive. State Historical Society of Missouri, Columbia.

Seton, Ernest Thompson. *The Book of Woodcraft.* Garden City, N.Y.: Garden City Publishing, 1912.

———. *Two Little Savages.* New York: Grosset & Dunlap, 1911.

Thorp, James. Scrapbooks. 8 vols. Western Historical Manuscript Collection, Columbia, Mo.

I. In the Bottoms

Hamilton, Henry W. *The Aftermath of War: Experiences of a Quaker Relief Officer on the Polish-Russian Border, 1923–1924.* Dayton: Morningside House, 1982.

———. Correspondence and miscellaneous papers. Western Historical Manuscript Collection, Columbia, Mo.

Hamilton, T. M. Correspondence and miscellaneous papers. Western Historical Manuscript Collection, Columbia, Mo.

James, Edwin. *Account of an Expedition from Pittsburgh to the Rocky Mountains, Performed in the Years 1819, 1820, . . . under the Command of Maj. S. H. Long.* Vol 14 of *Early Western Travels, 1748–1846,* edited by Reuben Gold Thwaites. Cleveland: Arthur H. Clark Company, 1905.

Lewis, Meriwether, and William Clark. *The History of the Expedition under the Command of Lewis and Clark.* 3 vols. Edited by Elliott Coues. 1893. Reprint, New York: Dover Publications, 1965.

Mattes, Merrill J. *The Missouri Valley: A Student's Guide to Localized History.* Localized History Series, ed. Clifford L. Lord. New York: Teachers College Press of Columbia University, 1971.

Mutel, Cornelia F. *Fragile Giants: A Natural History of the Loess Hills.* Iowa City: University of Iowa Press, 1989.

II. Hanging Mart Rider

Dyer, Thomas G. "'A Most Unexampled Exhibition of Madness and Brutality': Judge Lynch in Saline County, Missouri, 1859." *Missouri Historical Review* 89 (1995): 269–89, 367–83.

Foote, Shelby. *The Civil War: A Narrative.* 3 vols. New York: Random House, 1958–1974.

Hamilton, Jean Tyree. *Abel J. Vanmeter, His Park and His Diary.* Marshall, Mo.: Friends of Arrow Rock, n.d.

———. *Arrow Rock: Where Wheels Started West.* Marshall, Mo.: Friends of Arrow Rock, 1963.

Hamilton, T. M., et al. *Indian Trade Guns.* Union City, Tenn.: Pioneer Press, 1982.

Hurt, R. Douglas. "Planters and Slavery in Little Dixie." *Missouri Historical Review* 88 (1994): 397–415.

III. The Missouri Princess and Petit Missouri

Anderson, Bernice G. *Indian Sleep Man Tales: Authentic Legends of the Otoe Tribe.* New York: Greenwich House, 1984.

Berry, J. B. "The Missouri Indians." *Southwestern Social Science Quarterly* 17 (1936): 113–24.

Bossu, Jean-Bernard. *Travels in the Interior of North America, 1751–1762.* Edited and translated by Seymour Feiler. Norman: University of Oklahoma Press, 1962.

Bray, Robert F. "Bourgmond's Fort d'Orleans and the Missouri Indians." *Missouri Historical Review* 75 (1980): 1–32.

———. "European Trade Goods from the Utz Site and the Search for Fort Orleans." *Missouri Archaeologist* 39 (1978): 1–76.

Chapman, Carl H., and Eleanor F. Chapman. *Indians and Archaeology of Missouri.* Rev. ed. Columbia: University of Missouri Press, 1983.

Dorsey, J. Owen. "Migration of the Siouan Tribes." *American Naturalist* 20 (March 1886): 210–22.

———. "Siouan Sociology: A Posthumous Paper." *Annual Report of the U.S. Bureau of American Ethnology* 15 (1893–1894): 213–44.

———. "A Study of Siouan Cults." *Seventh Annual Report of the Bureau of Ethnology for 1889–90* (1894): 351–544.

Dorsey, J. O., and C. Thomas. "Missouri." *Bulletin of the Bureau of American Ethnology* 10, 1 (1907): 911–12.

Draper, Lyman Copeland. "Notes Regarding the Boonslick." Draper Manuscript Collection, State Historical Society of Wisconsin.

Foley, William E. *The Genesis of Missouri: From Wilderness Outpost to Statehood.* Columbia: University of Missouri Press, 1989.

Foster, Lance M. *"Tanji na Che:* Recovering the Landscape of the Ioway." In *Recovering the Prairie,* edited by Robert F. Sayre. Madison: University of Wisconsin Press, 1999.

Gregg, Kate L. "The War of 1812 on the Missouri Frontier." *Missouri Historical Review* 33 (1938–1939), 3–22, 184–202, 326–48.

Hamilton, Henry W. *The Archaeology of Saline County, Missouri.* Museum of Anthropology and Missouri Archaeological Society, 1985.

———. "The Spiro Mound." *Missouri Archaeologist* 14 (1952).

———. *Tobacco Pipes of the Missouri Indians.* Columbia: Missouri Archaeological Society, memoir no. 5, 1967.

Hamilton, Henry W., Jean Tyree Hamilton, and Eleanor F. Chapman. *Spiro Mound Copper.* Columbia: Missouri Archaeological Society, memoir no. 11, 1974.

Hamilton, T. M. *Colonial Trade Guns.* Chadron, Nebr.: Fur Press, 1980.

———. *Native American Bows.* York, Penn.: George Shumway, 1972. Reprint, Columbia: Missouri Archaeological Society, special publication no. 5, 1982.

Henning, Dale R. "The Oneota Tradition." In *Archaeology on the Great Plains,* edited by W. Raymond Wood, 345–414. Lawrence: University of Kansas Press, 1998.

Hurt, R. Douglas. *Nathan Boone and the American Frontier.* Columbia: University of Missouri Press, 1998.

Lévi-Strauss, Claude. "Four Winnebago Myths: A Structural Sketch." In *Culture in History: Essays in Honor of Paul Radin,* edited by Stanley Diamond, 351–62. New York: Columbia University Press, 1960.

McGee, W. J. "The Siouan Indians: A Preliminary Sketch." *Annual Report of the U.S. Bureau of American Ethnology* 15 (1893–1894): 157–204.

Nasatir, A. P., ed. *Before Lewis and Clark: Documents Illustrating the History of the Missouri, 1785–1804.* 2 vols. St. Louis Historical Documents Foundation, 1952.

Norall, Frank. *Bourgmont, Explorer of the Missouri, 1698–1725.* Lincoln: University of Nebraska Press, 1988.

O'Brien, Michael J. *Paradigms of the Past: The Story of Missouri Archaeology.* Columbia: University of Missouri Press, 1996.

O'Brien, Michael J., and W. Raymond Wood. *The Prehistory of Missouri.* Columbia: University of Missouri Press, 1998.

"The Prehistoric and Historic Habitat of the Missouri and Oto Indians." In *Oto and Missouri Indians.* New York: Garland Publishing, 1974.

Radin, Paul. "Literary Aspects of Winnebago Mythology [with four tales]." *Journal of American Folk-Lore* 39 (January–March 1926): 18–52.

Shunatona, Richard. "Otoe Indian Lore." *Nebraska History* 5, 4 (October–December 1922): 60–64.

Smith, David Lee. *Folklore of the Winnebago Tribe.* Norman: University of Oklahoma Press, 1997.

Thomas, David Hurst, et al. *The Native Americans: An Illustrated History.* Edited by Betty Ballantine and Ian Ballantine. Atlanta: Turner Publishing, 1993.

Thomas, Davis, and Karin Ronnefeldt, eds. *People of the First Man: Life among the Plains Indians in Their Final Days of Glory.* New York: E. P. Dutton, 1976.

Tixier, Victor. *Tixier's Travels on the Osage Prairies.* Edited by John Francis McDermott. Translated by Albert J. Salvan. Norman: University of Oklahoma Press, 1940.

De Villiers, Baron Marc, et al. "Massacre of the Spanish Expedition of the Missouri (August 11, 1720)." Translated by Addison E. Sheldon. *Nebraska History* 6, 1 (January–March 1923): 1–31.

Waters, Frank. *Book of the Hopi.* New York: Penguin Books, 1977.

Woodress, James Leslie. *Willa Cather: Her Life and Art.* New York: Pegasus, 1970.

IV. Mother, Father, Farm

Berger, G. W. "Limiting Age of the Miami Mastodon by Thermolu-
minescence." *Current Research in the Pleistocene* 13 (1995): 77–78.

Dunnell, R. C., and T. M. Hamilton. "Age of the Miami Mastodon."
Current Research in the Pleistocene 12 (1995): 91–92.

Earley, Amber, et al. "Dating the Miami Mastodon." Poster for 1999
Society for American Archaeology, Chicago.

Hamilton, T. M. "The Miami Mastodon." *Missouri Archaeologist* 54
(1996): 79–88.